THE COSMIC WINTER

WE DEDICATE THIS BOOK TO
FRANCIS, TRISTAN, MARIE-ANNE AND BENEDICT
AND TO
BRUCE AND AILSA

THE COSMIC WINTER

VICTOR CLUBE
AND
BILL NAPIER

Basil Blackwell

Copyright © Victor Clube and Bill Napier 1990

First published 1990

Basil Blackwell Ltd
108 Cowley Road, Oxford, OX4 1JF, UK

Basil Blackwell, Inc.
3 Cambridge Center
Cambridge, Massachusetts 02142, USA

British Library Cataloguing in Publication Data

A CIP catalogue record for this book is available from
the British Library.

Library of Congress Cataloging in Publication Data
Clube, Victor.
The cosmic winter / Victor Clube and Bill Napier.
    p.   cm.
Includes bibliographical references.
    ISBN 0-631-16953-9
    1. Civilization–Extraterrestrial influences. 2. Glacial epoch.
3. Catastrophical, The. 4. Historical geology.   I. Napier, Bill.
II. Title.
CB156.C56   1990
001.9–dc20   89-18011 CIP

Extract from Thorkild Jacobsen, *The Treasures of Darkness*
reproduced by permission of Yale University Press,
copyright © 1976 by Yale University

Typeset in 11 on 13 pt Ehrhardt
by Joshua Associates Ltd, Oxford
Printed in Great Britain by
T.J. Press Ltd, Padstow, Cornwall

Contents

Acknowledgements

It is beyond the authors' power of recall to identify all those scientific colleagues who have wittingly or unwittingly contributed in the course of discussions to the development of the ideas expressed in this book. We are indebted to them nevertheless, and no less to those whose expertise lies in disciplines remote from our own. Our particular thanks are due to Maurice Pope who critically reviewed the whole manuscript, and to Peter James who crucially directed his attention our way, although the responsibility for what is said in the pages that follow must rest entirely with ourselves. Our thanks are also due to Angus Macdonald who not only provided substantial bibliographic help but also translated the entire volume of Radlof (see plate 12) from the original German; to Ruth Cannell for further translations from German; and to Marjorie Fretwell for drawing many of the diagrams. Acknowledgements for permission to reproduce individual illustrations are given in the plate and figure captions. Particular mention should be made of Brian Warner for his help in obtaining the Piazzi Smyth illustration of the zodiacal light (plate 6) and H. U. Keller for the photograph of the nucleus of Halley's Comet (plate 9(c)). Last but by no means least, it is a pleasure to thank the team from Basil Blackwell who produced *The Cosmic Winter*: our editors Romesh Vaitilingam and Mark Allin who responded graciously to our every whim; likewise Ginny Stroud-Lewis and Sophie Hartley who devoted much care and patience to the production of the illustrations; and finally, Tracy Traynor, Brigitte Lee and Graeme Leonard who brushed up our loose ends and introduced the general air of tidiness that the original manuscript lacked. As ever, we are indebted to our respective wives Moira Clube and Nancy Napier who have been continual sources of inspiration and encouragement as the present work progressed.

The priests of the mysteries tell us what they have been taught by the gods or mighty daemons, whereas the astronomers make plausible hypotheses from the harmony that they observe in the visible spheres.

Julianus, Emperor of Rome, AD 361-363

Prologue

It is 7.17 am, Eastern Standard Time, on the thirtieth of June 1994. Suddenly the power fails.

The basement centre of the Defense Communication Operations Unit adjoining the White House is plunged into darkness, and only the distant rumble of early morning traffic reminds the duty officer that another day has already begun. There have been power failures before, however, and there is no hint of impending crisis.

Almost immediately, an automatic switch brings in the emergency supply. But now the duty officer observes that something is wrong. Telephone links to the outside world have gone. These include the open links, via commercial cables, to NORAD's bunker a thousand feet under Cheyenne Peak in Colorado, and to the Offutt command post near Omaha.

Within seconds it is ascertained that neither the President nor Vice-President, both of whom are out of town, can be contacted: radio communications are blacked out, on all frequencies. Further, because of the telephone failure, none of the Presidential stand-ins scattered around town and country can be expected to call in to give their exact locations, as they would normally if there were a recognized national emergency. However, the Secretary of Defense, third in the chain of command, is at breakfast somewhere in the White House.

It has taken 45 seconds to confirm the general breakdown of communications; it takes another 45 seconds to locate the Secretary of Defense and inform him of the problem; and it is another 90 seconds before the Secretary finds himself in the cramped under-ground Situation Room along with his advisors. He is aware that the travel time of a missile from a hostile submarine off the Atlantic coast to Washington is just over fifteen minutes; three of these minutes have gone.

Emergency procedures for transferring authority are now in hand

and the Secretary of Defense becomes in effect the new President. The 'gold codes', which would enable him to initiate a nuclear response, are placed before him. The black SIOP books with their theoretical scenarios are produced. The 'Mystic' and 'Nationwide' communication systems are successfully activated utilizing deeply buried, protected cables. He now has control of the National Command Authority, the channel of communication for execution of a nuclear strike; and another two minutes have gone.

The underground cables to Cheyenne, Offutt and other SAC bases are now open with secure links to the radar lines covering the northern approaches. None of them – Dew Line, the 55th Parallel or Pine Tree – is functioning properly, however; their screens are strongly upset with many spurious signals: there seem to be huge scattering modes in the ionosphere. Satellite communication with military forces abroad is impaired. No information is coming in from the big synchronous DSP satellites over South America, the central Pacific or the Indian Ocean; nothing is arriving from the relay stations near Aurora in Colorado or Pine Gap in Australia. And the 'hot line' is dead. On the other hand in the minutes before the confusion they had reported no anomalous infrared emissions and hence no evidence of heat from the exhausts of rising missiles. At the moment all that can be said is that there is a very unusual atmospheric disturbance, that telephone lines are dead and that there are widespread power failures. The disturbed atmosphere might just be some strange effect due to sunspots or the like, but the telephone and power failures are ominous. On such a scale, the only plausible explanation is an electromagnetic pulse caused by nuclear fireballs.

Eight minutes into the emergency – if it is an emergency – the Secretary issues a number of precautionary orders. His deputy is ordered upstairs to await the arrival of a helicopter from Quantico Marine Corps base about 30 miles away. This will transport the deputy to the waiting 'Nightwatch', a Boeing 747 which functions as an airborne command post. Another helicopter is despatched to retrieve the President. A 'flash alert' is sent to Hawaii, to get a second command post aloft, but the message goes unacknowledged. SAC missile silos and bomber bases are put on full alert.

Nine minutes on, information begins to flood in from sensors and antennae in orbit, at sea and scattered round the Earth. Much of it is reassuring: there are no patterns of activity on land, sea or air consistent with hostile intentions, no evacuations of barracks or the like are in progress. But now something deeply disturbing is coming in from one of the DSP satellites. A television screen shows a brilliant patch over south-east Nevada, bright enough to saturate the lead sulphide cells in the satellite. Intense heat is radiating from an area of about ten thousand square kilometres. Distant ground sensors are reporting large tremors emanating from the same area – unnecessarily, as the Situation Room has begun to sway and vibrate, and loud rumblings are coming up from deep under the ground. Then, from Offutt, comes devastating news. Air Force pilots are reporting a huge explosion in the desert area close to Boulder City. The city itself has gone, reduced to rubble. Las Vegas and other towns within 100 miles are enveloped in huge firestorms. The Hoover Dam has disintegrated. A column of dust and rubble has been sucked up to great altitude and a mushroom cloud is spreading outwards. Huge amounts of smoke are blanketing the state of Arizona and spreading into New Mexico and Mexico itself. The Secretary and his advisors can see it all on their television screens.

The Secretary now has a maximum of five minutes in which to discuss and evaluate the information, make any decisions and implement them. He is informed that the damage corresponds to explosions amounting to at least twenty megatons, and must have been caused by more than one bomb. The possibility of accidental detonations is quickly ruled out, because of its technical improbability and the magnitude of the explosion. Further, he is told that both ground and air bursts must have been involved to disrupt both cable and ionospheric communications. The conclusion seems unavoidable that for some reason the Soviets have targeted bombs on to American territory having somehow circumvented military radars. Perhaps armed satellites with low radar cross-section have suddenly been diverted downwards; or perhaps submarine-launched cruise missiles have been fired; or perhaps bombs had already been smuggled into the area. But why? The desert is a bizarre choice of target, Boulder, Las Vegas and the Hoover Dam of low strategic value. No sense can be made of the attack. Perhaps, it is suggested, they were chosen *because* they have little value. The attack may be a prelude to some major military adventure and may be a warning, a 'keep out' notice pitched at a level low enough that the risks of nuclear response are not justified and yet high enough to demonstrate deadly earnest. Or, the impacts may be intended to disrupt radar and communications for reasons which can only be guessed at.

Whatever the merits of such speculations, the Secretary cannot responsibly rule out the possibility that further bombs may follow, completely jamming command and control over nuclear weapons, or even that Washington is an imminent target whose destruction might for example be awaiting only the positioning of armed satellites, already in orbit, over the city, or the arrival of cruise missiles. Twelve minutes into the emergency, the Secretary's deputy is still on the White House lawn waiting for a helicopter, there is no communication with the President and the only information still coming in confirms that towns over several thousand square miles have been destroyed in nuclear blasts. He has no time left and must come to an immediate decision.

He could initiate the dreaded 'Major Attack Option'. This is quickly ruled out because it is beyond reason, leading only to mutual annihilation. He could choose to do nothing. But it is pointed out that in the present situation paralysis could also lead to holocaust. Whether the National Command Authority is 'decapitated' in the next few minutes depends on whether further missiles are about to land. If they are, then all control over the counter-strike force may be lost and, with bombs falling on the USA, a full-scale attack on the Soviet Union will, following standing instructions, be unleashed by individual submarine commanders, and by the one-star generals aboard their Cover All airborne command posts. It is therefore vital to forestall any further nuclear attacks by calling the bluff, that is by retaliating at a similar level. All the options are dangerous. The least dangerous is an immediate 'controlled response'.

The discussions are interrupted by confused reports of a storm of missiles, a cataract of fire pouring in over the states of the western seaboard and Canada. But the reports go unconfirmed; again, communications are failing over the whole of the United States.

It is 7.30 am, Eastern Standard Time, on the thirtieth of June. In the revamped over-the-horizon radar base near Gomel in Byelo-Russia officers are horrified to see, through the confusion of the still-disturbed ionosphere, a dozen missile tracks appearing on their screens. Peacekeepers are rising over the plains of Kansas. . . .

Even as they look, however, another unheralded missile is encountering the atmosphere high over the Bernese Oberland. Coming in at sixty thousand miles an hour, a hundred kilometres above Interlaken, the missile is gathering around itself a narrow skin of compressed air whose temperature surges up to about half a million degrees. Waves ripple within the hot air-turned-plasma; atoms, scattering off the waves, reach huge speeds, collide violently, and are stripped to their nuclei; electrons, torn from the atoms, accelerate violently and radiate fiercely at all wavelengths. For the most part this fierce emission is invisible, hard ultraviolet light or soft X-rays, and the atmosphere absorbs these at altitude, 50 to 90 kilometres up. Only a tiny fraction of the radiation reaches the ground as visible light. Even so, Europe is lit up as if with a flashbulb: survivors from Ireland to Austria and from Denmark to Italy later describe 'a bluish white light, too bright for the naked eye', or 'a fire brighter than the Sun', or 'a thunderbolt ... blinding in its intensity'. Approaching from the south east, the missile passes on a long, shallow trajectory, throwing moving shadows over the Jungfrau, Berne and Basle. It crosses the sky in a few seconds, leaving a wake of hot, luminous debris and air surrounded by a rapidly expanding red trail. Ten seconds after its first encounter with the atmosphere, the missile has penetrated the ionosphere, passed through the stratosphere and is moving rapidly into the lower atmosphere.

Here at an altitude between 10 and 15 kilometres, the missile hits significantly denser air and suddenly disintegrates. It never reaches the ground therefore and instead unloads its stored-up power, 200 megatons of impact energy, into the air over Louvain, a medium-sized town in central Belgium. The missile vaporizes in a third of a second. It is now, momentarily, an incandescent cylinder a few kilometres long and a few hundred metres across. The temperature within the cylinder is over 100,000° C. The pressure for an instant reaches some tens of thousands of tons per square inch; the missile blows up into a ball of fire radiating X-rays of huge intensity. These are absorbed within a few metres by the surrounding air; but, in doing so, the envelope of atmospheric gas acquires their energy. Simultaneously, hot, compressed material is thrusting rapidly outwards from the point of disintegration and snowploughing the air ahead of it. A blast wave develops; the fireball expands. From the first trace of incandescent tail high over Interlaken, to the beginnings of the fireball over Louvain, less than eleven seconds have elapsed.

When the fireball has reached about four miles in diameter the shock wave breaks away and races ahead of it. The corresponding blast brings a wind speed of over a thousand miles an hour. Survival at this distance is impossible; Louvain, directly under the fireball, disappears in less than a second. The fireball meanwhile inflates and soars rapidly upwards for several kilometres until, in the stratosphere, it flattens out and takes the classical mushroom shape. The cloud is seen from Copenhagen to Florence, from Edinburgh to Budapest. And over the whole of Europe, the ground shakes and buildings sway dangerously. Trains are derailed in southern England and beneath the Channel. The mushroom cloud is quickly broken up by jet streams at about 60 kilometres altitude and is subsequently dispersed around the globe.

Within 25 kilometres of Louvain, all buildings are reduced to rubble. A few girder bridges survive due to accidental cancellation of pressure waves from above and below. The shock wave reaches Brussels to the west and Liege to the east. The buildings within these cities collapse, domino-like, as the wave sweeps through; and all traces of roads are immediately erased by rubble. Beyond 25 kilometres, some structures survive, although few houses remain intact. Throughout Antwerp, for example, about 50 kilometres from ground zero, roofs are stripped bare of tiles and window frames and doors are blown in. Beyond this the damage becomes lighter: about 100 kilometres to the north the blast, sweeping through Eindhoven, blows tiles off roofs and shatters windows.

People out of doors and 25 kilometres from Louvain are hurled about 10 metres and are usually killed. Flying glass, though, is the greatest hazard. Shopping precincts and office complexes within 50 kilometres find themselves swept without warning by lethal blizzards of flying glass. Indeed, shards of glass account for most of the loss of life in these areas. Even 100 kilometres from the epicentre of the explosion, the blast is still strong enough to flatten forests.

Nuclear reactors within reach of the ground tremors and air waves are damaged, their coolant pipes snapped and containment vessels cracked. Some begin the process of meltdown. Already, swathes of radioactive debris are pouring into the atmosphere, mingling with the dense chemical smog, drifting into Germany, and destined for Poland and Scandinavia.

Immense destruction is caused by fire as well as blast. Serious flash burns from the thermal pulse of the fireball are sustained over an area of several thousand square kilometres, and virtually all exposed individuals within about 50 kilometres of ground zero receive third degree burns (charred skin) requiring labour-intensive hospital treatment, but in the hours and days following the impact it is impossible for medical services from outside to reach, let alone cope with, the seriously burned, and many of the injured die untreated from shock within twelve hours.

The intense heat from the rising fireball sets off fires at a range of 70 kilometres or more. Fanned by the updraft of air sucked in behind the rising ball, these merge into a mass fire, a single all-consuming conflagration. Such fire services as have survived are immobilized in a sea of debris, deprived of water and overwhelmed by the extent of the raging fire. The Sun eventually sets, but Europe is already dark, only the light from raging fires penetrates the smoke.

Sixteen hours after the event the Sun rises again, but it shines down only on a dense, choking smog created by blazing cities and petrochemical complexes. Aerial survey is impossible. Within a circle of radius 50 kilometres, the dead greatly outnumber the living. Even without the mass fire which is still blazing, penetration of the area is impossible as roads are strewn by debris and wrecked vehicles. It will be several days before the fires burn themselves out and the smog disperses enough to reveal the flattened cities and towns, villages and farms. Belgium, it will turn out, has been obliterated. And several weeks would need to pass before French, German and Dutch forces have the capacity to clear the roads, shelter the living and bury the dead. Belgium, however, will wait in vain . . .

For the Earth has encountered a cosmic swarm. Twelve hours later, the death-dealing fusillade is peppering the other side of the globe. Indeed, the bombardment continues for nearly a night and day, the planet rotating all the while, first presenting one face, then the other, until eventually it emerges on the far side of the swarm. Compared with the fireball over Hiroshima at the end of the Second World War, that over Boulder was a thousand times greater whilst that over Louvain was ten times bigger again, but they are just two of the several hundred pieces of grape-shot that meet the Earth, and they are trivial compared with the larger pieces of ordnance in train. These rarer missiles are each capable of unloading as much energy as would be unleashed in a major nuclear war, and in their case the atmosphere provides relatively little protection. Thus unlike their smaller counterparts, these larger bodies get through to the surface of the Earth, an ocean impact generating a tsunami, a land impact generating a small crater. There is another difference: each huge fireball is now too energetic to be properly contained by the atmosphere. As before, it ascends, laden with dust, but now it breaks through generating a vacuum in its wake which is filled by a vast stream of air. Thus, on the ground, once the initial blast has swept past, there is a counterflowing hurricane, itself sweeping up dust thrown up by the impact, and blowing it upwards after the rapidly receding fireball. Within minutes the fireball has reached an altitude of several hundred kilometres, in the high stratosphere. Here it stabilizes, and the dust begins to spill outwards over the top of the atmosphere.

Closer to the ground, the intense heat of impact from one large fireball has generated hundreds of fires over an area about the size of France. Essentially everything combustible along the line of sight of the rising fireball has caught fire. Loss of life over this area, already massive from the prompt effects of blast and heat, becomes almost complete when these fires merge into a single, all-consuming conflagration. About fifty million tons of smoke pour upwards, in dense plumes rising 10 kilometres into the atmosphere.

Within a few days, the wildfires are almost global in extent. Tens of millions of tons of fine dust lofted into the stratosphere, and a comparable amount of smoke in the lower atmosphere, have spread over the northern hemisphere and are beginning to cross into the southern one. At ground level, sunlight is blocked from reaching the ground, and all activity takes place in a black, choking smog. There is no question of mounting any rescue operation to save the devastated areas: the damage is so widespread that communications around the globe have now effectively ceased to exist; individual regions are virtually isolated; devastated populations are thrown back on their own resources; dozens of cities are reduced to smouldering rubble; river pollution is rife; and forests throughout the world are ablaze. Life as we know it is drawing to a close.

The cosmic encounter is over. Our planet has finally emerged from the swarm. Both continue, of course, along their predestined paths, the swarm merely a few missiles less, the Earth now charred and encased in a lingering veil of dust and smoke. With the destruction of cities and urban areas, the infrastructure of civilization has already gone; with the loss of sunlight, continental land temperatures have already plummeted to those of a Siberian winter; thick ice covers rivers and lakes; storms of unprecedented intensity rage along the continental margins; with animal and plant life devastated, farming and agriculture have already collapsed. Soon, the disrupted weather patterns will cause the continental land masses to be blanketed in a thick covering of snow.

After several months, the Sun begins to be seen through a hazy sky. When, eventually, the dust clears, the land masses of the northern hemisphere are covered in a snowfield which, reflecting sunlight back into space, has become permanent. The snowfield is added to each year, centimetres at a time. A thousand years on, North America and Europe are covered in ice sheets half a mile thick and the ocean level has dropped by about fifty metres. The Earth is locked into a new ice age. Mankind has survived, but the human population has crashed and society has dissolved into hungry marauding bands. A new ecological balance is being worked out on the blasted landscape: indeed the struggle for survival has only just begun. . . .

What is to be made of a future that turns out this way? First and foremost, of course, it must be established whether a true prognosis has been given; and if not in exact detail, at least in general terms. Next, should the future be as bleak as it appears, perhaps it must be recognized that very little can at present be done to remove personal risk except by fostering such stoical reserves as are necessary to face the slings and arrows of outrageous fortune. But what of culture, society and civilization? For these also appear to be at risk from the course of events we describe. Does one perhaps comfort oneself with the thought that not all would be lost since it is a reasonable expectation that some archaeologist, thousands of years hence, is bound to recover the essential achievements of our civilization, preserved in London, say, beneath a silted-up Thames? Alternatively, does one perhaps take strength from the known capabilities of the human race, on the basis, for example, that an engineer or physicist, also thousands of years hence, is bound once again to invent the internal combustion engine? Or does one picture a world peopled by then with a race of frankenstein monsters, indifferent to all but themselves? That is, should one envisage a swarm in the sky which is also the bearer of a cosmic imprint capable of affecting cellular life and marking out a new small branch in the evolutionary tree? Is civilization, indeed, simply a cul-de-sac, a diversion before a new dominant species – 'homo unsapiens' we might suppose – inherits the earth?

Science has no answer to these questions. Indeed, those who

might be expected to contribute to the answer are not at the moment interested. The environment in which our planet moves is supposed to be empty, and to pose little threat. We shall show, however, that the reality is different; that an unrecognized hazard is out there; and that at a stroke, civilization could be plunged into a new Dark Age.

The future shocks described herein are the logical consequence of new discoveries which have been emerging over the past few years. Some preliminary findings were described in an earlier book, *The Cosmic Serpent*, but the analysis has since been greatly extended, and we did not there deal with the implications for the future. That omission is here rectified. As we anticipated, some 'authorities' reacted to the book with outrage, and indeed the reader should be warned: much of what he may have regarded as established truth will in these pages also turn a somersault.

In June 793, 'fierce foreboding omens came over the land and wretchedly terrified the people', signs that are said to have been followed by great famine. On 25 June 1178, the Moon, more or less in line with the Earth, was apparently struck by a missile whose energy was ten times that of the combined nuclear arsenals of the world. On 30 June 1908, an object from space exploded above a remote area of Siberia with the energy of a large hydrogen bomb. More recently, during five days in late June 1975, an unexpected swarm of boulders the size of motor cars struck the Moon at a speed of 67,000 miles an hour. And then again, on 30 June 1994, there came an unheralded explosion with the strength of a 20-megaton bomb. . . .

Why late June? What is the nature of these events? And what is the actual threat they pose for mankind? Such are the questions we address in this book. For within these last few years, it has been found that there is a great swarm of cosmic debris circulating in a potentially dangerous orbit, exactly intersecting the Earth's orbit in June (and November) every few thousand years. More surprisingly, perhaps, it has been found that the evidence for these facts was in the past deliberately concealed. When the orbits exactly intersect however, there is a greatly increased chance of penetrating the core of the swarm, a correspondingly enhanced flow of fireballs reaching the Earth, and a greatly raised perception that the end of the world is nigh. This perception is liable to arise at other times as well, whenever fresh debris is formed, but deep penetrations occurred during the fourth millennium BC, again during the first millennium

BC, taking in at their close the time of Christ, and will likely take place yet again during the millennium to come.

Christian religion began appropriately enough therefore, with an apocalyptic vision of the past, but in the aftermath of the last deep penetrations, once the apparent danger had passed, truth was converted to mythology in the hands of a revisionist church and such prior knowledge of the swarm as existed, which now comes to us through the works of Plato and others, was later systematically suppressed. Subsequently the Christian vision of a permanent peace on Earth was by no means universally accepted, and it was to undergo several stages of 'enlightenment' before it culminated with our present secular version of history, to which science itself subscribes, perceiving little or no danger from the sky. The lack of danger is an illusion however, and the long arm of an early Christian delusion still has its effect. *The Cosmic Winter*, then, is a kaleidoscope of history and science, reviving the basis of an old and largely misunderstood pagan view of the world.

The idea of a terrible sanction hanging over mankind is not, of course, new. Armageddon has been widely feared in the past and it was a common belief that it would arrive with the present millennium. During the last thousand years, moreover, it has usually been the reforming church that revived the fear. But such ideas, whenever they have arisen, have always met with fierce opposition. Sometimes the proponents of such ideas escape to new found lands where in due course they meet opposition of a homegrown kind. In the United States for example, despite freedom of speech, old traditions of cosmic catastrophe have recurred from time to time, even in the present century, only to be confronted by pavlovian outrage from authorities. That being the case, it is perhaps ironic that elections in the United States are generally held in November following the tradition of an ancient convocation of tribes at that time of the year, which probably had its roots in a real fear of world-end as the Earth coincided with the swarm.

In Europe the millennium was finally dispensed with when an official 'providential' view of the world was developed as a counter to ideas sustained during the Reformation.[1] Indeed to hold anything like a contrary view at this time became something of a heresy and those who were given to rabble-rousing for fear of the millennium were roundly condemned. To the extent that a cosmic winter and Armageddon have aspects in common, therefore, authoritarian

outrage is nothing new. More to the point perhaps, the way in which
the British parliament handles its affairs seems to have its roots in
the condemnation of the heresy.[2] Thus the mother of parliaments,
once its present charge was assumed, began by turning against the
founding sect known as Ranters. Ranters rose to prominence in the
period immediately following the execution of the English king and
the foundation of the Protectorate. But they are heard of very little
these days and it is not commonly known that Ranters took their lead
in 1649 from one Gerrard Winstanley who was moved by 'supernatural illuminations', along with many others, to anticipate the
millennium.

In the event, the 'destroying angel' failed to materialize, the
ranting parliamentarians came to look foolish and it was only a
matter of time before the reins of power passed to others of a more
sophisticated persuasion whose vision of the world embodied the
kind of 'enlightenment' that has sustained western civilization ever
since. Enlightenment, of course, builds on the providential view and
treats the cosmos as a harmless backdrop to human affairs, a view of
the world which Academe now often regards as its business to
uphold and to which the counter-reformed Church and State are
only too glad to subscribe. Indeed it appears that repeated cosmic
stress – supernatural illuminations – have been deliberately programmed out of Christian theology and modern science, arguably
the two most influential contributions of western civilization to the
control and well-being of humanity.

As a result, we have now come to think of global catastrophe,
whether through nuclear war, ozone holes, the greenhouse effect or
whatever, as a prospect originating purely with ourselves; and
because of this, because we are faced with 'authorities' who never
look higher than the rooftops, the likely impact of the cosmos figures
hardly at all in national plans. Indeed, even those who do look higher
are mostly content simply to weave plausible hypotheses from the
harmony they observe in the visible spheres, and eschew any
reference to cosmic hazards in the process.

And if the overall climate of our globe should once again improve,
as it is doing during this century, and has done every few centuries
since the end of the last ice age, there may be only the dimmest
perception of an approaching nadir. We may be unaware that the
cosmos is simply delaying the next input of dusty debris, alarm,
destruction and death. A great illusion of cosmic security thus

envelopes mankind, one that the 'establishment' of Church, State and Academe do nothing to disturb. Persistence in such an illusion will do nothing to alleviate the next Dark Age when it arrives. But it is easily shattered: one simply has to look at the sky.

The outrage, then, springs from a singularly myopic stance which may now place the human species a little higher than the ostrich, awaiting the fate of the dinosaur.

I

THE LABYRINTH OF HISTORY

Plate 1. Destruction of Sodom and Gomorrah (from *The Signs of Heaven* by M. Bischoff).

I

Cataracts of Fire

Pallas rushed from the peaks of Heaven like the bright star sent by the son of crafty-counselled Chronos (as a sign either to sailors or the broad array of nations) from which many sparks proceed.

Homer, *The Iliad* [3]

Scarce had the old man ceased from praying when a peal of thunder was heard on the left and a star gliding from the heavens amid the darkness, rushed through space followed by a train of light; we saw the star suspended for a moment above the roof, brighten our home with its fires, then tracing out a brilliant course, disappear in the forests of Ida; then a long trail of flame illuminated us, and the place around reeked with the smell of sulphur. Overcome by these startling portents, my father arose, invoked the gods and worshipped the holy star.

Virgil, *The Aeneid* [4]

If the earth and sky had no starting point in time, why have no poets sung of feats before the Theban war and the tragedy of Troy? Why have so many heroic deeds recurrently dropped out of mind and found no shrine in lasting monuments of fame? The answer, I believe, is that *this world is newly made*: its origin is a recent event, not one of remote antiquity. That is why even now some arts are still being perfected: the process of development is still going on. . . . Alternatively you may believe that all these things existed before, but that the human race was wiped out by a burst of fiery heat or its cities were laid low by some great upheaval of the world. . . . All the more reason, then, to concede my point and admit that an end is coming to earth and sky. If the world was indeed shaken by such plagues and perils, then it needs only a more violent shock to make it collapse in universal ruin. . . . *There is no lack of external bodies to rally out of infinite space and blast [the world] with a turbulent tornado or inflict some other mortal disaster.* . . . It follows, then, that the doorway of death is not barred to the sky. . . . Legend tells of one occasion when fire got the upper hand. . . . The victory of fire, when earth felt its withering blast, occurred when the galloping steeds that draw the chariot of the Sun swept Phaethon from the true course, right out of the zone of ether and far over all the lands. Then the Father Almighty, in a fierce gust of anger, struck down the aspiring Phaethon with a sudden stroke of his thunderbolt, down out of the chariot to the earth. But the Sun intercepted the everlasting torch of the firmament in its fall, brought the trembling steeds back to the yoke from their stampede and, guiding them along their proper course, restored the universe to order: Such is the story as recited by the ancient bards of Greece, a story utterly rejected by true doctrine. *What may really lead to the triumph of fire is an increase in the accumulation of its particles out of infinite space.* Then comes the crisis: either its forces for some reason or other suffer a setback, or the world shrivels in its parching blasts and comes to an end.

Lucretius, De Rerum Natura[5]

God, whose dwelling is in the sky, shall roll up the heavens as a book is rolled, and the whole firmament in its varied forms shall fall on the divine earth and on the sea; and then shall flow a ceaseless cataract of raging fire and shall burn land and sea, and the firmament of heaven and the stars and creation itself it shall cast in one molten mass and clean dissolve. Then no more shall there be luminaries, twinkling orbs, no night, no. dawn ... no spring, no summer, no winter, no autumn.

Sybilline Oracles[6]

In the west a star shall shine, which they call a comet, a messenger to men of the sword, famine and death.

Sybilline Oracles[6]

Our eyesight is not able to pass through the middle of a celestial body to see through it to the things on the other side. But through a comet the things beyond are seen. ... Accordingly it is obvious that a comet is not a celestial body. ... Zeno judges that stars come together ... and from this union of light there comes into existence the image of a rather long star. Therefore some suppose that comets do not exist but that only the appearance of comets is rendered through ... the conjunction of stars clinging together. ... *Some say that comets have their own orbits and after fixed periods of time they come into men's view.* Others say that one should not give them the name of celestial bodies because they do not last long and are dissipated in a brief period of time. ... We see various kinds of fire conceived on high, sometimes the heavens blazing, sometimes 'long trails of flame

glowing white behind', sometimes great fiery bodies hurtling by. . . .
All men are amazed at those phenomena which carry sudden fire
down from on high, whether something flashes and disappears or the
atmosphere is compressed and forced into glowing, and it is taken as a
miracle. . . . Sometimes [the stars] do not wait for night but burst out
and shine well within daytime. . . . Why do they appear at a time not
their own? *It is generally agreed that stars exist even when hidden*.

Seneca, *Naturales Quaestiones*[7]

Dynasty Han, Reign Yuan-yan, Year 1, Month 4, Day Ding-you. At
the hour of *rifu*, the sky was cloudless. There was a rumbling like
thunder. A meteor with a head as big as a *fou*, and a length of some
ten-odd *zhang*, colour bright red and white, went south-eastward from
below the Sun. In all directions meteors, some as large as basins,
others as large as hens' eggs, brilliantly rained down. This only ceased
at evening twilight.

Ancient Chinese Observations of Meteor Showers[8]

Fierce, foreboding omens came over the land of Northumbria and
wretchedly terrified the people. There were excessive whirlwinds,
lightning storms and fiery dragons were seen flying in the sky. These
signs were followed by great famine . . .

Anglo-Saxon Chronicle, June 793[9]

The middle of the fourteenth century was a period of extraordinary
terror and disaster to Europe. Numerous portents, which sadly
frightened the people, were followed by a pestilence which threatened
to turn the continent into an unpeopled wilderness. For year after year
there were signs in the sky, on the earth, in the air, all indicative, as
men thought, of some terrible coming event. In 1337 a great comet
appeared in the heavens, its far-extending tail sowing deep dread in
the minds of the ignorant masses. During the three succeeding years
the land was visited by enormous flying armies of locusts, which
descended in myriads upon the fields, and left the shadow of famine in
their track. . . . What with famine, flood, fog, locust swarms, earthquakes, and the like, it is not surprising that many men deemed the
cup of the world's sins to be full, and the end of the kingdom of man to
be at hand. . . . An event followed that seemed to confirm this belief. A
pestilence broke out of such frightful virulence that it appeared indeed
as if man was to be swept from the earth. Men died in hundreds, in
thousands, in myriads, until in places there were scarcely enough
living to bury the dead, and these so maddened with fright that
dwellings, villages, towns, were deserted by all who were able to flee,

the dying and the dead being left their sole inhabitants. It was the pestilence called the 'Black Death', the most terrible visitation that Europe has ever known. . . . London lost one hundred thousand of its population; in all England a number estimated at from one third to one half of the entire population (then probably numbering from three to five millions) were swept into the grave. If we take Europe as a whole it is believed that fully a fourth of its inhabitants were carried away by this terrible scourge. For two years, the pestilence raged, 1348 and 1349. It broke out again in 1361–62, and once more in 1369. The mortality caused by the plague was only one of its disturbing consequences. The bonds of society were loosened; natural affection seemed to vanish; friend deserted friend, mothers even fled from their children . . . [others] seeing no hope of relief from human action, turned to God as their only refuge, and deemed it necessary to propitiate the Deity by extraordinary sacrifices and self-tortures. The flame of fanaticism, once started, spread rapidly and widely. Hundreds of men, and even boys, marched in companies through the roads and streets, carrying heavy torches, scourging their naked shoulders with knotted whips, which were often loaded with lead or iron, singing penitential hymns, parading in bands which bore banners and were distinguished by white hats with red crosses. Women as well as men took part in these fanatical exercises, marching about half-naked, whipping each other frightfully, flinging themselves on the earth in the public places of the towns . . .

The Romance of Reality [10]

Of course, the fashions and prejudices of the day will always colour our perception of the past. And from our supposedly enlightened perspective of the twentieth century, it may seem perfectly natural to regard Armageddon by divine dispensation as so much moonshine. We can hardly escape the fact, however, whatever its basis, that the dreaded expectation of fire from heaven is a pregnant component of our intellectual heritage (Plates 1, 2).

Indeed, the fear is apparently as ancient as civilization itself. A modern historian [11] for example, writing of the period five thousand years ago when man's great intellectual adventure seems to have begun, comments that Mesopotamian civilization grew up in an environment which was signally different from our own. As in Egypt, there were the same great cosmic rhythms, the change of seasons, the unwavering sweep of the Sun, Moon and stars, but also '*an unusual element in the sky of force and violence*'. As if to imply that this element may not also have been present in Egypt – which is, as we

shall see, questionable – this historian went on to construct a somewhat contrived antithesis between the erratic behaviour of the Tigris
and the Euphrates, on the one hand, and the supposedly more predictable behaviour of the Nile on the other. But he also noted the
scorching winds which smother men in dust, and threaten to
suffocate him; in addition, the torrential rains which turn firm
ground into a sea of mud and rob man of his freedom of movement.
The abiding impression, the historian seeks to assure us, is of man
standing amidst these great powers, sensing his weakness and realiz-
ing with dread that he is caught in the interplay of giant forces in
Nature. His mood is therefore tense, and his lack of strength creates
for him an awareness of tragic potentialities.

To explain, for instance, that certain atmospheric changes broke a
drought and brought rain, we are informed that the Sumerians and
Babylonians observed much the same facts as us but experienced
them in terms of interventions by the gigantic bird Imdugud which
came to their rescue:

'It covered the sky with the black storm clouds of its wings and devoured the
Bull of Heaven, whose hot breath had scorched the crops.' In telling such a
myth, the ancients did not intend to provide entertainment. Neither did they
seek, in a detached way and without ulterior motives, for intelligible
explanations of the natural phenomena. They were recounting events in
which they were involved to the extent of their very existence. They
experienced, directly, a conflict of powers, one hostile to the harvest upon
which they depended, the other frightening but beneficial: the thunderstorm
reprieved them in the nick of time by defeating and utterly destroying the
drought.[12]

Only by supposing the ancients dramatized the mundane, then, can
the historian apparently come to terms with the intensity of the
experiences of Nature which gave rise to the ancient Mesopotamian
mood and thus account for their disturbing notion of the cosmos: a
cosmos of predictable order rather than anarchy, it seems, but one,
nonetheless, that was far from safe or reassuring. What emerges,
then, from detailed studies of the written records and inscriptions
left by Sumerians and Babylonians is an attitude to nature which
obliged them to contemplate the universe with a sense of foreboding
hardly short of paralysis. It is this apparently strange attitude, sustained for millennia, that seems to have led to the full complement of
diviners, soothsayers, exorcists and astrologers which was in due
course to be found in the great temples of Babylonia.

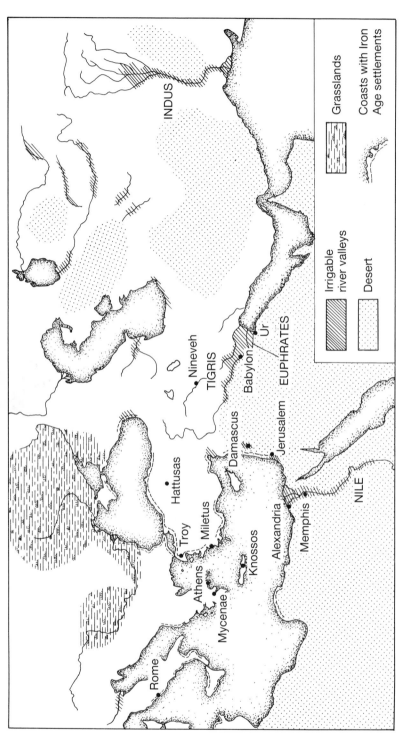

Figure 1. Beginnings of civilization. Map including the major areas, apart from the Chinese plains, in which we have evidence of the origin of agriculture and the building of cities in remote antiquity (adapted from an original illustration due to J. D. Bernal: *Science in History*).

We know from early Babylonian references that their most ancient manual of astrology, an encyclopedia of omen texts entitled 'Enuma Anu Enlil' ('Concerning the beginning of Anu and Enlil'), covers a period that stretches back well into the third millennium BC. The omen texts were essentially lists of correlated phenomena, dire terrestrial consequences of repeated astronomical events attributable to gods. Thus, in late Sumerian as in Old Babylonian writings, flood and deluge were sent by the gods, along with equally catastrophic visitations of plague, drought and famine, and there can be little doubt that the omens were used for the purposes of understanding natural phenomena and predicting the likely future course of events. To the extent, moreover, that there was a clear astronomical association with such phenomena as the Flood, there are hints that the astronomical visitations may have been regular as well as harmful, for in the Sumerian version of the Flood, Ziusudra, counterpart of the Old Babylonian Ut-Napishtim and the Biblical Noah, is represented as having an ability to forewarn and of resorting to divinatory practices; while Emmeduranki of Sippur, one of the kings of this age, anticipating events, is alleged to have obtained from the gods the skills and insignia of divination. We can be fairly sure, in fact, that the earliest form of astronomy to be practised was based on an interpretation of omens associated with regular, repeatable phenomena; and there is no question that these omens were derived from strange celestial portents, unusual atmospheric conditions or violent storms, each manifesting either the good will or malevolence of the gods concerning some known or unknown event in the future. It is possible of course, and even likely, that ordinary agricultural, calendrical and navigational requirements of astronomy were recognized as well. However we are left in no doubt at all that it was these mysterious signs from heaven and man's fear of their possible consequences that were largely responsible for fostering man's careful observation of the sky.

Five thousand years later and by an extremely tortuous path, this practice of skywatching has gradually evolved and as a result, man has now been led to his current 'scientific' understanding of the universe. This being the case, however, our modern view of the universe, indeed science generally, appears to be grounded in activities that seem to owe more to desperation than contemplation, more to fear than reason; a state of affairs with which scholars are not very comfortable. After all, if men of culture and privilege in another generation could build cosmology out of irrationality, out of a fear apparently rooted in imagination, how far can one trust the pronouncements of their modern-day counterparts? It is out of such concern – that their modern-day counterparts might not be trusted – that a modern savant[13] hastens to assure us that an archaeologist many years hence, who chances to unearth some twentieth-century newspaper, would naturally be led to conclude that we too regulated our lives in accordance with the predictions of astrologers. He is arguing, obviously, that cultured sections of the population among early civilizations would not have been affected by astrology and that astrological fatalism was always the device of charlatans, not proper astronomers, for the edification of morons who could afford to pay liberally for indulging such a cult. It is an attractive idea: it gets science off the hook, and it leaves the ancient astrologer to his strange devices! The idea, however, is without foundation: indeed it is wholly wrong. The fact is that rulers and ruled alike in ancient Mesopotamia had absolute faith in astrologers and that it was the advice of astrologers above all which was sought when it came to affairs of state.

On a straightforward interpretation then, putting aside our twentieth-century fashions and prejudices, it would seem that the sky as understood by the Babylonians harboured dangers which determined the destiny of nations. Indeed the level of conditioning to this idea was so complete that the assumption itself would never have been questioned. This is of course a striking fact; and why intelligent men, let alone the remainder of the population, should in the past have formulated a view of the universe so threatening and so unlike that which presents itself to the casual observer today has never been satisfactorily resolved in modern times – unless, of course, we are happy with the assumption that common sense and human courage are recent acquisitions.

There are two possible ways of explaining this hiatus in understanding. Either the ancient sky behaved in some manner which *was* truly different from the present day, or modern man is indeed significantly more matter-of-fact and enlightened than his ancient counterpart. The second alternative happens to be the one that is chosen! Our modern savant, for example, goes on to say of the historical development of his subject that it 'shows how the conception of a continuous interference and control of the universe by some external power gave way to a nobler and higher conception of an inherent natural order. A mechanical view of the universe replaced the older and cruder view, which was quite consistent with arbitrary and vindictive acts on the part of a petty-minded ruler. The modern astronomer pursues his research on the assumption that such acts do not occur and without this assumption, he could not with any degree of confidence continue his investigations.'

To its credit, the statement is as clear an expression of fundamental principle as one could hope to find and it certainly describes accurately enough the lofty turn of mind that is so often cultivated amongst the learned today. But how can we be sure that it is not based on an insecure modern preconception? How can we be sure, in other words, that the sky never changes, that it never produces a cataract of raging fire?

At first sight this question still seems bizarre. After all, there were professional sky-watchers in the past and they have left us their records. Surely it is inconceivable that the disinterested scholars who analyse these records could have missed anything so dramatic as a cataract of raging fire? Indeed it could be argued that the very fact that we are conditioned to believe the sky was *not* dangerous is evidence enough that it was not: after all, the experts who have examined the evidence thoroughly must have arrived at this conclusion in a balanced way. On the face of it, therefore, this is not an issue that can suddenly become a topic for renewed speculation: one is surely dealing with reasonably established historical facts.

But of course, history is not all a collection of hard facts.[14] Like science, history is a matter of inference; and records may in principle mislead unless an accurate picture of what they describe is also at hand. The question that is posed therefore is whether the records left by the ancient sky-watchers are so transparent as to their meaning that no misunderstanding can possibly exist. Is it absolutely certain, then, what the Babylonian astrologers had in mind?

Until fairly recently, this has not seemed a particularly interesting question to ask. It has generally been assumed that the first thoughts of any real value about the universe are those handed down to us by the Greeks. However this assumption has been seriously questioned by Otto Neugebauer,[15] the distinguished historian of mathematics and astronomy, and it is his conclusion that Babylonian skywatching should not be regarded as irrational, even though its nature and purpose did eventually undergo a profound change. This finding has not as yet had very much influence on the generality of scientific historical studies. Nevertheless, it is Neugebauer's thesis that a quite significant scientific revolution was engineered by the Greeks around the fourth century BC and that little progress can be made in understanding ancient science and history until the precise nature of this intellectual upheaval has been fully explored. On the face of it, a scientific methodology developed and perfected over a thousand years earlier in Hammurabi's Babylon was put aside by astronomers of this later period in favour of new analytical techniques along with new understandings of nature. In essence, the practice and theory of 'omen astrology', or judicial astrology as it is sometimes called, gave way at the very foundation of western civilization to the new-found theory and practice of 'horoscopic astrology'.

We shall be examining various aspects of this development at greater length in a later chapter but for the moment we note two points. On the one hand, the Greeks are now clearly seen as having taken over certain recent advances in the mathematical methods of the Babylonians and fashioning them in new and unexpected directions. On the other hand, arising from this step, the Greeks can now be seen also as having abandoned at this time the prevailing view of the cosmos and cultivating an entirely new understanding of how it works.

On the mathematical side, it was the Babylonians, as we now know, who developed the precise algebraic techniques that enabled the combined effects of various superposed periodicities in the motion of heavenly bodies to be calculated. These techniques were essential in determining the apparent position of the Moon and in setting up an accurate calendar based on its motion.[16] However, it is perfectly clear that the Babylonians also knew, as we do now, that such calculations could be done without reference to any physical model of the Moon's behaviour. They understood that they were merely projecting into the future, with greater precision than would have been possible without mathematics, such trends and correlations as had been detected amongst past measurements. But it was the Greeks who took over the principles of this algebra and applied them to geometrical models: they suggested that the positions of the planets in the sky could be explained as the combined effects of various superposed *circular* motions, and by these means they came to invent the theory of epicycles.

No harm would have come of this otherwise laudable computing device had it not then happened that Greek philosophers took to thinking the epicycles were physically real. There emerged the idea that planets, with the Sun and the Moon, were somehow attached to invisible crystalline spheres interlocked by a complicated and equally invisible system of revolving gears. Soon, by the force of their own logic, the Greeks were convincing themselves that nothing out of the sky would penetrate these crystalline spheres to harm the Earth below.[17]

Astronomers and scientists generally have, of course, long realized that the epicyclic system was an illusion but in directing these particular Greek thoughts to the intellectual dustbin, it has been all too easy to hang them on the Babylonian forebears of the Greeks as well. In fact, there is no good reason for supposing the Babylonians were similarly captivated or that it affected their view of the cosmos. Indeed, as Neugebauer explains, all the evidence indicates that the Babylonians, unlike the Greeks, were quite straightforward in their dealings with astronomy – the universe may have been frightening but they sought to describe it much as they saw it, in simple commonsense terms. To them, there would have been nothing odd about things out of the sky harming the Earth. If, then, the Babylonians appear to confront real things in the sky, as the historians would have us believe, and if, also, we have good reason to believe that the astrologers were straightforward and scientific, why can we not settle the issue by examining the astrologers' actual records and discovering the precise nature of the sky they described?

The ancient city of Babylon first came to prominence early in the second millennium BC, as Semitic immigrants gradually merged with and supplanted the indigenous Sumerian population.[18] Sumerian culture was not lost however and it was the founding of new schools of language and theology which apparently created the atmosphere of general learning that emerged in Babylon and which then provided the milieu in which the discipline of mathematics was able to originate. There also emerged a powerful new ecclesiastical school beholden to the god Marduk. This deity, whilst not altogether displacing Anu, the highest of the ancient Sumerian gods, now achieved great power and completely replaced earlier animal divinities such as Enlil (probably a bull) and Ea (probably a ram) who were also of high rank in the Sumerian pantheon. Apart from their ecclesiastical duties however, the priests of Marduk were also astronomers and it is they who seem to have initiated the practice and theory of omen astrology. Under their guidance, celestial phenomena came to be used for predicting the imminent future of the country and its government, particularly the king. From the appearance or non-appearance of 'planetary' bodies, conclusions were drawn concerning the invasion of enemies from the east or the west, and in addition the arrival of floods and storms, but nothing like a horoscope based on a constellation at the moment of birth has ever been found. In other words, Babylonian astrology right from the start is seen as being much better compared with weather prediction from phenomena observed in the skies than it is with the casting of horoscopes. Indeed it is now hardly to be doubted that this omen astrology was soundly based on the concept of observed irregularities in nature which were indicative of other disturbances to come, and in this respect it differs not at all from the practice of modern science.

But whilst the principles underlying the methodology seem reasonable, the same is not generally said of the actual observations. Thus the specific and fundamental irregularities that the astronomer-priests of Marduk were in the habit of recording were the emergence of new-born lambs and the appearance of the liver or other internal parts of lifeless sheep! Thousands upon thousands of clay tablets have been unearthed on which were kept the updated accounts relating to the delivery of these animals or their fleeces. Such observations immediately conjure up a rather mundane or distasteful image: the astrologers seem to be more at home in the market place, or to indulge in strange sacrificial practices that have little to do with the honest pursuit of knowledge. However, we should not forget that astronomer-priests were also the spokesmen on earth for generations of leaders who claimed their divine right to rule from sky gods who, as shepherd-kings, were said to have descended with their flocks from Heaven at the time of the Flood. It may not be coincidental therefore that the recurrence of a catastrophic event like the Flood should have been anticipated on the basis of deviant behaviour by sheep, something which must presumably have preceded the Flood itself. For if each sheep was in fact some kind of *celestial* object which was in the habit of going into predictable decline before fragmenting into a number of bright offspring under circumstances that were not without danger to our planet, the astronomer-priests' custom of recording the arrival of sheep and of examining the state of their various internal parts might not seem so strange. Once again, of course, the reader will expect that this possibility has been carefully examined and found wanting, so it will be something of a surprise to learn that the actual Babylonian character for a wandering star, *lubat*, literally means stray sheep! In addition, the observations often relate to a particular ring in the sky (the zodiac) whose literal meaning is that of a fence or stream enclosing a sheep meadow. The records we interpret, in other words, do not distinguish between sheep and stars! The actual imagery indeed is of sheep that are scuttling around a pen in the sky, developing ailments that have some adverse consequences so far as our livelihood down here is concerned. Why then are the cuneiform scholars so perverse as not to take this imagery at face value? Why do they insist on understandings of the data which are either sacrilegious or mundane? Presumably, it is because they have been unable to relate any known astronomical phenomenon to any image of sheep scuttling around the sky that is also threatening.

Fortunately, it is possible to resolve the question of what the astronomer-priests were doing by examining the way in which omen astrology developed. Thus, although the political power of Babylon was in due course eclipsed, the city maintained a thousand-year-old uninterrupted tradition of culture and learning which eventually, in the first millennium BC, attracted the admiration of much younger states. Indeed from the seventh century onwards, we find Persians, Jews and Greeks alike, all apparently drawn to Babylon by the magnet of 'Chaldean' science, sparking off something akin to a renaissance in the cosmopolitan atmosphere of easy communication and competition that then seemed to emerge. However it should be kept in mind that a revival of interest in the sky can also be sparked off by events in the sky itself. Whatever the setting, Babylonian mathematics and astrology experienced a great revival which continued into the so-called Seleucid period, and it is perhaps no surprise to find visitors such as Zarathustra and Pythagoras, like Abraham much earlier, being proclaimed in their own lands as the inventors of all science and the creators of astrology and number-wisdom, each then asserting himself as the oldest and hence the teacher of mankind.[19]

In later chapters, we shall be tracing the thread of science along its labyrinthine path through ancient Greece and the empires of Macedon and Rome to the present day. Pliny, for example, will

Plate 2. Final Judgement. Based on a sixteenth-century fresco at the Dionysian Monastery, Mount Athos in Northern Greece (Roger Bargue). Note the highly realistic portrait of a meteor shower associated with seismic disturbance and divine control.

inform us that 'fleece' was still a commonplace term for comets at the beginning of this era (chapter 6). But more to the point, paths east also carried the theory and practice of Babylonian astrology through Persia and India to China where it experienced yet another renaissance towards the end of the first millennium AD[20]. And here, at last, there can be little doubt about what the astrologers were doing.[21] Their preoccupation was first and foremost with the detection of 'guest stars' and with the enumeration of fireballs, the former now known to be largely comets and the latter, those extra-bright meteors, now known to be largely their offspring. But, as Schafer has recently remarked in relation to the acknowledged status of these practices, 'where the historian of Chinese science [now] tends to be concerned with the accuracy of such ratios and measurements as the obliquity of the ecliptic and the length of the year, the astrologers of the T'ang [were] more concerned with the benign or alarming visions that signalled to them from the black vault'.

Even today there are sky-watchers around the world who continue in the Babylonian tradition of Chinese astrologers and occasionally observe the diminishing tail of a fading comet, or record the arrival of shooting stars. But one thing is certain: they do not commonly reflect on what the ancients may have been doing when they examined the dwindling entrails or registered a count in the presence of incident sheep. The thought that a dead comet might metamorphose into a dangerous swarm of 'hidden stars' is one that the fashions and prejudices of the present day do not allow. Logic requires us to keep our shepherd-kings in the ancient sky and regard them as a myth!

The problem then is one of interpretation. The Babylonian astrologer did indeed describe the danger in the sky. Furthermore, throughout history, observers have made the link between catastrophe down here and danger up there. Cuneiform scholars are not of course unaware of the ambiguities that arise out of the fact that the earliest sky-watchers were shepherds. There must therefore be a deeper conviction to which they subscribe, namely that the sky does not, after all, harbour any danger. In the chapters that follow, we shall attempt to trace how this conviction came about and consider whether it has any basis in fact.

2

Forces of Evil

I⊤ is in the nature of things that bodies of knowledge tend to be slotted into models, or theoretical pictures of reality. It often happens that new facts come along which do not fit perfectly into old ideas, but in practice small modifications to the overall structure usually ensure that it remains intact. It occasionally happens, however, that inconsistent facts so accumulate that they cannot be accommodated within an existing framework; things then break down, and sometimes it is necessary to have a full-scale revolution in thinking. In this situation old facts remain valid but they are understood in a quite different way, with new patterns of thought. To the participants, truth itself often seems to do a somersault although, of course, it does not: it is simply the perception of the truth that is being discarded.[22]

It is, however, difficult to judge when an accumulation of cracks in a structure can be repaired, and when they are signalling that collapse is imminent. Indeed, scientific reputations are made and lost on such judgements. It is easy in retrospect to marvel at establishment obtuseness, but a given situation is rarely clear-cut at the time, and for every successful scientific revolution there are many more which fail or never get off the ground. But if truth often advances in this fitful way, it must follow that erroneous models of reality are held for periods of time which may be quite long, before the weight of evidence destroys them. It also follows that there may well be ideas which are widely or universally held to be true today, which are in fact false. There are many historical precedents – geocentrism and Biblical creationism are two familiar examples – which show that humanity is perfectly capable of holding on to the wrong preconceived model for thousands of years: human reasoning on its own is very fallible and experience shows that it is liable to go more and more astray unless it is continually checked against new observations or experiments. It is, then, usually new evidence which eventually destroys erroneous preconceptions; but when this happens, the old evidence too has to be re-thought.

A revolution is under way in the Earth sciences. Terrestrial catastrophism, the idea that the evolution of life and even fundamental geological processes are controlled by sudden inputs of material from space, has become an important if controversial concept in these fields within the last few years (see chapters 14 and 15). The evidence is partly in the rocks, but it also derives in large part from new astronomical observations. The standard view advocated for a century or more by experts in geology and biology, that the Earth evolves in splendid isolation from its surroundings, is proving to be wrong. What we are doing in this book is taking the same astronomical discoveries and applying them to timescales of historical rather than geological interest. That is, we show how the same astronomical evidence which is now leading to new insights in the earth sciences also tells us that cataracts of fire must have taken place within the last few thousand years. We consider how historical evidence is better fitted to this new catastrophist framework.

The reader must not be deceived, however. We also deal in this book with matters at the boundary of human knowledge and which, for one reason or another, are accepted as understood within a quite different framework. Cataracts of fire, if contemplated at all by the expert, are considered to be absurd. The question we are inviting the reader to consider therefore is whether the facts are already beyond the level where the standard point of view can still absorb them. The Babylonians, for example, have told us quite plainly that they were afraid of things in the sky and, as we have seen, they described what they observed in terms that leave little doubt as to the appearance of the danger they perceived. And yet, the modern interpreter elects to discount this knowledge, believing that what was observed must have been imagined. We shall argue indeed that this is not the only instance and that large bodies of written evidence are now ignored or cursorily dismissed because they do not fit into a preconceived twentieth-century framework, which was, after all, constructed before the new discoveries in Earth science and the new astronomical evidence came along. Indeed, given our opening story, there is another dimension to the discussion: can the all-too-common scholarly approach, of holding on to cherished views by the fingernails until forced off by the sheer weight of evidence, be *morally* justified? What justification can there be for sustaining an unshakeable conviction that may put civilization at risk?

Between seven and five thousand years ago, our planet was experiencing a relatively benign climate. The conditions seem to have played some part in causing communities in several different areas of Asia and North Africa to evolve the first urban economies (figure 1). Thus by 3000 BC, rather similar great civilizations outside China had emerged on the alluvial soils of three major river systems: the Tigris–Euphrates in Mesopotamia, the Nile in Egypt and along the Indus Valley in an area that corresponds to much of modern Pakistan. The advance of these civilizations evidently depended in part on the production of sufficient food to support not only the section of the population involved in such work but also the specialist classes engaged in craft industries and administration. The necessary surplus was apparently achieved through the transfer of farming practice in the highland zones to the alluvial plains of these particularly extensive river valleys which could be exploited, through irrigation, to produce high crop yields.

Around 3000 BC however, and lasting for about two centuries, the terrestrial climate deteriorated quite markedly on a global scale.[23] Rainfall increased and the average temperature declined. The northernmost treeline in Canada and Northern Europe for example fell back several hundred miles (figure 2); and this in spite of the proliferation of forests elsewhere. Mountain glaciers increased their range. During this period, too, major flooding occurred in Mesopotamia and Egypt; the former evident in the silt excavated beneath ancient citadels and the latter revealed by the location of temples relative to the fluctuating Nile. Elsewhere large-scale changes in vegetation and ground cover were taking place, including those brought on by extensive forest fires supposedly due to clearances undertaken in a period of presumed agricultural prosperity.[24] It is a paradox of any global climatic recession that whilst some areas experience a worsening of general conditions, others may enjoy an amelioration of the environment.[25] In general though, we can be sure there was change: for some a turn for the better, for others, a period of prolonged strife.

Remarkably, the same epoch brings clear evidence of a surge in civilization; new skills, the appearance of writing and the establishment of a professional class, all coinciding essentially with the

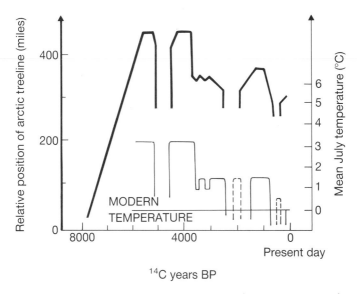

Figure 2. Mean summer temperatures (global average relative to its present day value) based on the movements of the Arctic treeline during the last 7,000 years (reproduced from *The Little Ice Age* by J. Grove). The sustained amelioration during the period 6000–3500 BC corresponds to the recognized climatic optimum of the Holocene period, whilst the broken lines indicate possible brief ameliorations due to a reduced incidence of atmospheric dust veils, as indicated by acidity peaks in Greenland ice cores and frosted tree rings.

start of the historical era. Moreover it may be inferred from the various new state enterprises that were undertaken after this time, for example the construction of pyramids and the implementation of large irrigation schemes, that the population was motivated beyond any previous expectation. Judgements of what was required and levels of application in human effort emerged that completely transcended earlier practice. The question naturally arises whether these dramatic developments had anything to do with the change of climate. We shall ask also whether the pace could have been set by an unexpected overpowering force, related to the change in climate, which simultaneously created a sense of urgency and no little apprehension.

Neither of these lines of research seems to have been seriously pursued by modern archaeologists and historians. Most of these experts assume we are dealing with a coincidence and that the physical change of climate around 3000 BC could at best have had only a minor influence on the advance of civilization. A century ago, in the aftermath of the industrial revolution, indeed before the physical evidence of climate change was known, it seemed natural to explain the parallel rise of the great 'hydraulic empires' in terms of the diffusion of knowledge and lucky breaks in practical know-how. Prehistory was therefore envisaged as a succession of more advanced technological states – stone, bronze, iron and so on – each phase gaining more leisure wherein to plan subsequent more rapid advances of mankind. Something of the influence of Darwin's theory of evolution can be seen in this way of looking at prehistory.

However the scheme of never-ending technological evolution, so alluring to the nineteenth- and twentieth-century mind, began to look rather vulnerable when it was realized that some Europeans were extracting and processing copper well before the fourth millennium BC. It was clear then to the experts that if the idea of evolution was not to be abandoned, technology would have to become something of an incidental theme. The notion emerged therefore that separate communities would evolve in parallel because of some innate natural law of human behaviour. Given similar initial conditions, it was assumed, communities would develop similar internal 'socio-economic pressures' driving them to similar states of civilization. Such pressure would cause mankind to advance from a primitive state of savagery through nomadic pastoralism and barbarism to settled agriculture and mature civilization. The same pressures would gather homesteads into villages, merge villages into city states, cause urban civilizations to grow and monolingual empires to spread, until the species inherited the Earth. The technological fashions of the mid-nineteenth century have been replaced by the sociological ones of the mid-twentieth.[26]

A line of thinking has developed therefore in which it now seems most natural to subordinate the ultimate course of history to man's civilizing urge. By following the technological fashions of the nineteenth century and the sociological ones of the twentieth, in fact, the archaeologist and the historian may have contrived to keep environmental factors at bay. If one studies the Wessex culture of Early Bronze Age Britain, for example, it may be to discover whether it arose locally during trade with the continent, or whether it was imposed by continental invaders on a subject population. The answer provides insight perhaps into the processes by which different races of people gain dominance over one another and advance the fate of nations. But by attaching importance to such patterns in human behaviour as may then emerge, and regarding them as an end in themselves, without introducing any allowance for the differing physical circumstances and changes of environment, there is always the risk of imposing patterns on the course of history that bear no relation to its underlying motivation. Sociological formulae are rather limited in their scope therefore. Indeed, if information is limited, and this is usually the case, the issues are often intractable and virtually impossible to resolve. Nevertheless, if it seems there is no other course to follow, the programme of enquiry may continue to be justified to the exclusion of others. The result is that one learns only very slowly, if at all, whether a productive line of enquiry is being pursued. In the present instance, therefore, it may be asked whether archaeologists and historians have committed themselves to a line of enquiry that is potentially interminable. With such a restrictive programme, apparently self-imposed, it seems there may be only a slender chance of returning to the fundamental issue at stake: that is, whether the *symptoms* of civilization, man's technological and sociological feats, the rise and fall of nations, reflect the incidence of some deeper historical force.

To get back to this question, however, we are bound to study what the ancients themselves said. So far we have simply considered the state of the Mesopotamian mind and the activities of astrologers several millennia ago, but perhaps one of the most surprising aspects of all that went on is the significance of the role the ancients attach to gods. In Egypt also, at the start of the third millennium BC, people were attributing their welfare and misfortune to divine beings of comparatively recent provenance. It behoves us, perhaps, to see now whether anything more can be learned about these gods. One might imagine, for example, in view of the climatic changes at the time, that purely symbolic 'weather gods' were invented as a convenient mode of expression and that these subsequently got out of hand! However this would certainly be to understate the enormous scope of the gods in Egyptian eyes, even if it were tolerable on other grounds.

Thus it was from about 3000 BC onwards that the dynastic pharaohs took for granted their infusion of godly power derived from an ancestral deity.[27] And with divine names like Scorpion, Catfish, Fighter, Serpent and Killer, for example, it is hardly to be doubted that the early kings looked upon their gods as quite a bellicose bunch. Eventually, the feudal sparring between Egyptian chieftains gave way to a more omnipotent pharaoh and a more regal style of existence. Horus, a falcon-god, became the sky-god associated *par excellence* with kingship: essentially a benevolent figure, he provided security and continuity going back to the very foundation of the new Egyptian state. It is no denial, of course, of any contribution of climatic change to possible conditions of strife if we follow the Egyptians themselves awhile and scrutinize carefully the role of Horus, whom they took to be their accepted saviour.

The multifarious nature of Egyptian mythology arises from the fact that different city-states claimed the special protection of different gods. With the amalgamation of cities, the god of one city was liable to be identified with the god of another. It was customary however for each god to have its own family and relations, many of them being in animal form, while in due course, the more significant ones tended to become humanized. Despite the immediate impression, therefore, the Egyptians have analogues of the divine shepherd with his sheep, and the mythologies have a monotheistic flavour from a very early stage: the single, very remote god Ptah is one such example associated with generations of subsequent deities. According to the earliest cosmic legends, Osiris and Seth were first-generation paternal gods of considerable importance. Respectively black and red skinned, they also brought into existence 'two lands'. These lands, Blackland and Redland, were apparently the two main areas of a supposedly huge flat world. The contemporary geographical world centred on Egypt seems moreover to have been incorporated into Blackland, the latter being bounded by a continuous enclosure which separated it from Redland. This enclosure, constructed by Osiris, clearly lay in the cosmic realm however since it also marked the zodiacal path of the celestial barque in which the Sun-god and a further complement of sky gods made their daily journey.

It was implicit therefore in this early Egyptian conception of the world that the flat Earth was more or less continuous with a flat cosmos. Egyptian temples were even modelled on the arrangement, the enclosure being represented by a rectangular colonnade and wall whose interior face was decorated with a waving stream surmounted by a starry sky. The enclosure was in fact synonymous with a river or stream whilst Redland beyond, literally outside, was frequently identified as an ocean or sea. Thus, another aspect of the imagery connected with Blackland and Redland was their supposed transformation from or emergence out of the primordial waters of chaos, which were also in some sense associated with the birth of Osiris and Seth.

Within the enclosure and at one end was the so-called Island of Creation, the possible repository of a meteoritic stone, as at Mecca, whilst at the other, on either side of the main entrance, were two huge pylons, a representation of features which were apparently thought to extend out of the cosmic enclosure into cosmic space, in an arrangement which seems to have been sufficiently part of the accepted scheme of things to have been later subsumed in places of worship as pairs of towers or minarets. This is an important aspect of cosmic structure to which we shall return at a later stage. For the moment, we simply note that Osiris was believed to have been murdered by Seth who then dismembered his body and distributed its parts around the enclosure. Isis, the devoted sister and wife of Osiris, was later responsible for putting her husband's body together again, and thus resurrected, posthumously conceiving a child by him. The child, Horus, was raised to manhood by his mother and eventually avenged his father's death by repeatedly defeating Seth in battle.

To the Egyptians, especially during the Middle Kingdom, this recurring conflict epitomized past and future triumphs of the forces of good over the forces of evil, and they clearly identified Horus as the lord and protector of Egypt. Implicitly, there was an understanding that the forces of evil might sometimes make incursions into the territory belonging to the forces of good under Horus's protection, and there are references throughout Egyptian history to threats from these foreign peoples of the Sea. The collapse of the Middle Kingdom (c.1650 BC) may indeed be directly attributable to their influence. We read for example of 'a blast from God' which left Egypt in a state of 'dire affliction' without a sovereign upon the throne. 'Rulers of foreign lands' called the Hyksos then took over the country, apparently without a struggle, burning the cities and the temples of God. They were said to have Seth as their god and they are definitely presented as barbarous and ruthless overlords, though archaeologists and historians can find little or no material evidence from the period of their rule to support such a claim. To this day, there is no physical indication of invaders from any country with which the Hyksos may be identified. It seems they may as well have brought brief destruction from the sky! Seen in this light, Horus and

Seth and their cosmic setting do not seem unreasonable and one can understand perhaps why they should have been of such lasting significance for Egyptian life. We are led to conclude that the cosmic menace in the Mesopotamian sky, whatever it was, was observed in Egypt as well.

Indeed, wherever one looks in the contemporary ancient world, one comes across creation myths[28] referring to celestial giants who engaged one another in battle and who once dominated the terrestrial scene. This battle was likewise a recurring one which usually ended in victory by the forces of good. The victory over Seth by Horus has its counterpart for example in the defeat of the ancient Babylonian Tiamat by Marduk, the Greek Cyclops by Titans amongst whom Zeus was eventually to be regarded as the most prominent, the conquest of the Hebrew Behemoth by Leviathan, the contest with the Dragon that ended in victory for St. George and the overcoming of Satan by Yahweh, to name but a few (see table 1). No more than a cursory examination of these ancient myths is necessary to show that the forces of evil always had strong celestial overtones (see chapter 13), so the Mesopotamian civilization was far from being alone in its intense fear of the sky. Indeed it is difficult to escape the impression that specific agencies were present in the ancient sky, one of which was seen as a potential threat to man whilst the other seems eventually to have been regarded as a saviour. The question that has to be resolved therefore is whether there is any

Table 1 Sky gods in perpetual battle

Source	Benevolent force	Malevolent force	Displaced father-figure
Egypt	Horus	Seth	Osiris
Greece	Zeus	Typhon	Chronos
Babylon	Marduk	Tiamat	Enlil
Syria	Baal	Yam	*
Iran	Ohrmazd	Ahriman	*
Scandinavia	Thor	Odin	*
Medieval	Michael	Devil	*
Medieval	George	Dragon	*
Hebrew	Yahweh	Satan	*

It is characteristic of the recurrent conflict that the forces of evil eventually succumb to the forces of good, and in those cases where no father-figure is clearly identified, it is possible that a very ancient name has been preserved in that of the later benevolent force.

material basis for a dualism in the sky, one aspect plainly harmless and the other a clearly perceived danger.

It is interesting to note for example that Osiris is shown in ancient illustrations wearing a long white cloak. He is said by the Egyptians to have brought with him the inspiration for civilization and agriculture. Eventually however, he lost his former glory and died, though his rejuvenated spirit was said to have survived and lived in his successor. Each pharaoh, symbolized by Horus during life, thus became Osiris on death. The descent by Osiris into the underworld followed by his journey to the Place of Ascension, otherwise the Island of Creation, was symbolized by a formal procession leading to the final resting place, a pyramid, the latter in some sense representing a heavenly abode as real as the cosmic enclosure separating Blackland from Redland. Obviously there is a problem in identifying where symbolism ends and reality begins but one may speculate whether a regularly revealed object in the sky with the appearance of a long white cloak can break up and lose its former glory whilst building in space a temporary island and a temporary closed path from outside which the Earth may continue to be menaced.

In this connection, it is clearly important to note that the traditions relating to Typhon in Greek myth, who corresponds to Seth, the opponent of Osiris in battle, also have rather definite cosmic connections. Typhon is specifically referred to as a kind of comet by several classical authorities and in the first century after Christ, Plutarch[29] wrote:

as for the dimmed and shattered power of Typhon, though it is at its last gasp and in its final death-throes, the Egyptians still appease and soothe it with certain feasts of offerings. Yet again, every now and then, at certain festivals they humiliate it dreadfully and treat it most despitefully – even to rolling red-skinned men in the mud and driving an ass over a precipice (as the Koptos folk) – because Typhon was born with his skin red and ass-like.

The red, ass-like skin of a dimmed and shattered god seems distinctly obscure to the twentieth-century mind, and one might seek to dismiss it as an incomprehensible piece of symbolism. However we shall anticipate later discussion (chapter 12) by noting that as a description of a once-great but near-defunct comet it is entirely appropriate.

Naturally, if such happenings were observed in the sky, one can appreciate why ancient myths from all over the world, especially those with a cosmological flavour, should have acquired similar underlying themes and many details in common. Furthermore, if the happenings were responsible in some way for assaults on the Earth by cataracts of fire leading to the breakdown of law and order, we can expect some overall coherence in the pattern of world history: we can expect the simultaneous collapse of empires, populations on the move at much the same time, conflicts to be forced and new ideas to emerge, dominated perhaps by a deep fear of what the sky has in store. This, strangely, *is* the pattern of world history (e.g. see table 2) and many a scholarly reputation has crumbled already in the search for its underlying cause. One cannot, of course, immediately presume on this basis that an interaction of the kind under discussion must therefore have happened. What we can say, however, is that the evidence is more consistent with the proposition than we might otherwise have been disposed to believe.

To the extent that cataracts of fire of cometary origin may have been a general source of terror in the past and it is natural to expect ritual responses that would then be preserved by desperate people seeking to appease apparently uncontrollable cosmic forces, we can also expect that the orbital characteristics of any particular comet involved might be reflected in very ancient social practices which have by this time lost their meaning. Fire festivals for example, involving bonfires, processions with blazing torches, tumbling burning wheels down hills, hurling lighted discs into the air and many other variations on similar themes, are in fact commonplace throughout the world and it is significant that these tend to occur

---

Table 2 (*opposite*): The climate of the last 10,000 years appears to be divided by relatively sharp global recessions (shaded) into distinct periods (sub-atlantic etc.) whose differing average conditions correlate with the frequency of meteoric inputs (x) in the range of sub-cometary mass. Thus it may be that the most intense meteoric inputs correspond with little ice-ages (not all shown), capable in some instances of generating a dark age. Microscopic carbon particles and charred plants are detected in basal layers of peat laid down during the most conspicuous climatic recessions (*) and are due to extensive forest fires, often remote from civilization, suggesting an extreme natural cause (see text) rather than human activity (commonly assumed). The overall pattern of civilization's development is certainly more complex than the table suggests but climatic recessions, characterized by glacial advance, ocean fall, lowered tree-lines, peat formation and floral as well as faunal change, including human migration, appear also to be associated with higher degrees of human deprivation and conflict from which new national groupings emerge to forge new empires: the arrows reflect general advances through city-states, nation-states and empire to a world civilization which is not generally achieved. Coupled with these developments are changing cosmological paradigms reflecting the perceived significance of meteoric (M) and planetary (P) phenomena as well as the level of danger induced by the astronomical environment. The simple sketch of the evolution of British civilization in the final column relates to the discussion in chapter 7.

Table 2   Emergent civilization, implicity meteor activity and climate

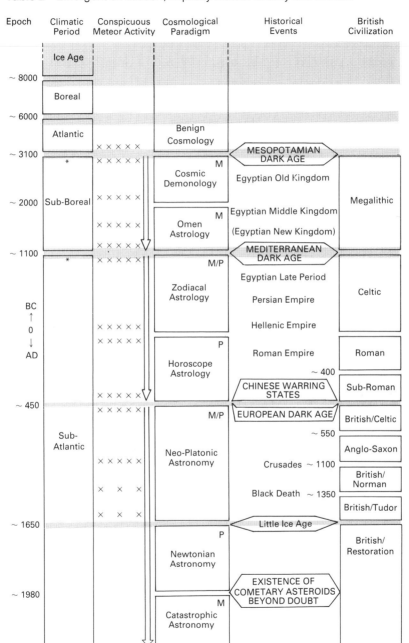

Epoch	Climatic Period	Conspicuous Meteor Activity	Cosmological Paradigm	Historical Events	British Civilization
	Ice Age				
~ 8000					
	Boreal				
~ 6000					
	Atlantic		Benign Cosmology		
~ 3100		× × × × ×		MESOPOTAMIAN DARK AGE	
	*	× × × × ×	M Cosmic Demonology	Egyptian Old Kingdom	
~ 2000	Sub-Boreal	× × × × ×			Megalithic
		× × × × ×	M Omen Astrology	Egyptian Middle Kingdom (Egyptian New Kingdom)	
~ 1100		× × × × ×		MEDITERRANEAN DARK AGE	
	*	× × × × ×	M/P Zodiacal Astrology	Egyptian Late Period Persian Empire	Celtic
BC ↑ 0 ↓ AD		× × × × × × × × × ×	P Horoscope Astrology	Hellenic Empire Roman Empire ~ 400	Roman
		× × × × ×		CHINESE WARRING STATES	Sub-Roman
~ 450		× × × × ×	M/P	EUROPEAN DARK AGE	British/Celtic
	Sub-Atlantic		Neo-Platonic Astronomy	~ 550	Anglo-Saxon
		× × × × ×		Crusades ~ 1100	British/ Norman
		× × ×		Black Death ~ 1350	British/Tudor
~ 1650		× × ×	P	Little Ice Age	British/ Restoration
			Newtonian Astronomy		
~ 1980			M Catastrophic Astronomy	EXISTENCE OF COMETARY ASTEROIDS BEYOND DOUBT	

most often during early November (Hallowe'en) and mid-June, the beginning of the former month being also commonly regarded in the past as the start of the calendar year. Whilst neither date has any conspicuous agricultural significance and the former bears no obvious relationship to the seasonal location of the Sun, there is a definite association between these times of the year and inter-sections by the Earth with the orbit of a particularly significant comet. We will return to this comet in the pages that follow (e.g. chapter 13) but for the moment, we again note connections which are more direct than we might otherwise have been disposed to believe.

Mythology and ancient ritural then, whether Egyptian, Mesopo-tamian or of some other source not considered here, point rather strongly to a persistent conflict with and within the ancient sky. Modern interpreters, historians and archaeologists, have nearly always chosen to give the evidence a figurative interpretation, believing the references to deities can only be symbolic. We shall not argue the matter any further for the moment; nations and empires have declined in modern times for reasons which are clearly more political than cosmic, so there is no question of a unique one-to-one relationship. Our purpose at this stage is simply to demonstrate that there are reasonable grounds for doubt.

But if the alternative picture is correct, the periods of stability in history which normally capture our attention lose something of their significance. It is the times when nations fall into chaos and ruin that mark the real turning points in civilization. Between crises, we might expect a semblance of order even if it takes a century or two to achieve, whilst a threat from the sky, even over a decade and even if it never materializes, may be sufficient to cause a delicately balanced social equilibrium to be lost for ever. Egyptian history, for example, is characterized by periods of warring city states and lack of central cohesion, even amounting to chaos and dark age, out of which there arose long periods of stability dominated by powerful, magnificent and generally stable kingdoms. These kingdoms, closely paralleled by periods of high civilization in Mesopotamia (table 3), appear to have collapsed suddenly in circumstances that are indeed not properly understood, each time initiating a new period. The approx-imate dates for the start of these periods, based on conventional chronology, are 3100, 2200, 1650 and 1250 BC.[30] The first corresponds to the close of the prehistoric era, the second to the close of the Old Kingdom, the third when the Middle Kingdom was abruptly

Table 3 Egyptian and Mesopotamian civilizations in conventional chronological sequence[a]

Dates BC	Principal Egyptian periods	Principal Mesopotamian dynasties	Dates BC
2800—2155	Old Kingdom	Akkad	2630—2150
2052—1600	Middle Kingdom	Amurru (Old Babylonian)	2110—1600
1554—1072	New Kingdom	Cassite	1530—1160
570—332	Late Kingdom	Chaldean/Persian	626—336
	(664—525 Sais Dynasty)	(Neo-Babylonian)	
332—	Hellenic Empire	Seleucid Empire	332—

[a] The cultural highspots in the development of Egyptian and Mesopotamian civilization are apparently correlated, though it also has to be kept in mind that the chronology, being largely geared to Manetho's Egyptian king-lists and the insecurely calibrated Sothic calendar, is probably only approximate.

terminated by the arrival of the Hyksos invaders, and the fourth in the interregnum years following Rameses II when Egypt temporarily succumbed to attacks by the Sea-Peoples. Though civilization then continued in reduced circumstances, the collapse in this final instance was apparently so devastating that Egypt never again succeeded in attaining its former glory. It was a collapse that did not go unrecognized elsewhere.

3

The Heraclids

It was not just in the valleys and deltas of great rivers like the Tigris and the Euphrates, or the Indus and the Nile that civilization first took an upward turn. The great urban civilizations coming at the start of the third millennium BC may have been the most conspicuous but further to the west, the inhabitants of the Aegean were also shaking off their Stone Age ways. Less impressive at first perhaps but no less accomplished, these people were also to reach a high level of artistic and technological achievement and by the beginning of the second millennium BC the Minoan civilization, centred on the island of Crete, had not only developed commercial and cultural links with the other great civilizations but had established itself as the dominant influence in the Mediterranean basin.

Relatively little is known of the Minoan contacts in other directions at this time, though Indo-European settlers penetrating from the north and east were already reaching the nearby Greek mainland. It used to be thought that these settlers spoke the same Aryan language from which are descended the Latin, Sanskrit, Celtic and Teutonic tongues, but the evidence indicates only that by 1600 BC, these settlers had combined with the population already present and had laid the foundations for a new Greek-speaking Mycenean culture on the mainland in a state of seemingly peaceful co-existence with their more powerful Minoan neighbours to the south across the sea.

The Greeks of classical antiquity[31] are indeed believed to have been drawn from two sources, representing something of a balance between an earlier indigenous population, which may have spoken a Semitic language, and an immigrant wave of peoples, from lands to the north and east, who spoke languages belonging to the Indo-European family. Legend seems to point to two ancient provinces of Greece, Achaea and Doris, where immigrant and earlier Aegean stocks were apparently superimposed. However the archaeological evidence suggests the latter was a comparatively late settlement around 1100 BC, and there is no indication that it was on any significant scale. Indeed it is generally accepted now that the principal Indo-European strain goes back to at least 2000 BC and that the feeling of racial unity that developed among the loosely organized groups, living in the narrow valleys between the mountain ranges and long gulfs which characterize the Greek archipelago, owed as much to the indigenous maritime people as it did to any superior warrior-caste invaders. Of course the Indo-European invaders could have been mainly responsible for the cohesion of the nation by imposing their feudal, aristocratic ways upon an otherwise subservient people; but it is also clear now that the Greek language was of local origin and that the sense of nationhood within the population could equally well be of considerable antiquity.

Thus the version of the Minoan script known as Linear B, which was used at the Cretan capital Knossos in the final period of its prosperity, is known to be an early Greek dialect identical with that current in Mycenean Greece at the time. This seems to indicate that Greek and Linear B reflect a language that was already fairly widespread in the Aegean by the mid-second millennium BC. The apparent predecessor of Linear B, namely Linear A, which was used during the period of the second Cretan palaces (1625–1450 BC), appeared around 1800 BC but has not yet been deciphered. On the other hand, in the period of the first Cretan palaces (1800–1625 BC), Linear A exists alongside the Cretan hieroglyphic script, traditionally so-called with reference to Egyptian hieroglyphics. Another offshoot of Linear A, Cypro-Minoan, developed during the sixteenth century BC though in circumstances which are not altogether clear. The evidence points, in fact, to a powerful Minoan civilization whose origins may well go back to the third millennium BC and which was eventually extinguished towards the close of the second millennium BC.

The upshot of all this is that archaeologists and historians have been able to piece together a fairly comprehensible picture of what went on in the area prior to 1500 BC. The achievement has merely served to throw into greater relief their increasing bewilderment at the subsequent course of Minoan and Mycenean history. For, in a relatively short period of time – no more than a few centuries – despite having reached a very high level of organization and affluence, both the Minoan and Mycenean civilizations dramatically collapsed. Indeed, they disappeared in what appear to have been curiously similar though separate circumstances: the island Minoan civilization, including its Aegean satellite colonies, around 1450 BC; and the more extensive mainland Mycenean dominion around 1200 BC.

It has been suggested that the Minoan civilization may have been destroyed by the Bronze Age eruption of the volcano on the island of Thera less than a hundred miles north of Crete, and the southernmost of the Cyclades. This was indeed a major event. Volcanic ash was scattered over a wide area of the eastern Mediterranean including the Cyclades and most of Crete: we can gain some idea of its violence from a comparison of Thera's caldera with that of Krakatoa. The latter is six times smaller yet in 1883, the corresponding explosion sent a huge tsunami through the narrow straits between Java and Sumatra drowning 36,000 people and destroying at least two hundred villages. However the topographies differ in each case, and there is no certainty that the Thera explosion brought the Minoan civilization to a close: the latter continued beyond the radiocarbon date of Thera (around 1500 BC) and seems to have disappeared under quite separate circumstances.

Although the *relative* dating leading to this conclusion, based on pottery styles above and below the ash deposits on Thera, is comparatively secure, it should be kept in mind that the *absolute* dating is less certain. The chronological yardstick to which all events in the ancient Near East are currently synchronized as far as possible, is the Sothic calendar of ancient Egypt. This is based on a sprinkling of astronomical observations set in the framework of king-lists which come to us through Manetho.[32] As we shall see, the final collapse of Bronze Age civilization in the Aegean extended throughout the Near East and produced a cultural hiatus that may have lasted for nearly five hundred years. The partitioning of states, giving rise to dynastic overlap, is therefore possible and remains a potential source of error in absolute chronology: it seems clear that we should still keep an open mind as to exactly when the Minoan and Mycenean empires finally disappeared.[33] However, none of this makes any difference to the substantiality of the events.

On both occasions, the countryside and the towns appear to have been devastated. As far as one can tell, the decline in civilization and the deterioration of the environment were more or less contemporaneous. Most Mediterranean areas were far more wooded and fertile during the Bronze Age than they are today; the present semi-barren state bears witness to a combination of destructive factors during the period in question involving extensive deforestation and virtual annihilation of the vegetation cover which caused the topsoil thereafter to be relentlessly eroded. It is even likely that the country-side and towns were destroyed on such a vast scale that many of the inhabitants were obliged to migrate, leaving only a small remnant after 1100 BC to populate and once again cultivate the land.

Be that as it may, the city buildings, whether they were the earlier unprotected palace towns of Crete like Knossos, Phaestos or Gournia, or the massively fortified citadels of mainland Greece such as Pylos, Mycenae or Tiryns, were devastated as if by earthquake or fire within two periods of time which were probably no longer than about fifty years, leaving the inhabitants on both occasions seriously weakened and considerably depleted. Despite its extent, which embraced the whole of Crete, the first of these occasions was probably on a lesser scale. Indeed, it was the collapse of the Minoan civilization that apparently provided the Mycenean empire with its principal opportunity for overseas expansion and for two-and-a-half centuries subsequently, its hegemony steadily grew until it embraced settlements as far afield as Egypt, Cyprus, Palestine, Syria, Troy, Cilicia, Sicily, southern Italy as well as Crete and the Aegean islands.

And yet, in due course, this civilization too was completely over-whelmed. This time, however, the whole area was left virtually devoid of culture for the subsequent five hundred years: essentially, a dark age was created. Very little of the Mycenean civilization survived and the whole nation evidently fell into a state of poverty, illiteracy and ignorance. The population was not only severely depleted, it descended to such a primitive level of existence that the skills of writing and architecture were no longer practised, indeed were largely forgotten. There is a continuous tradition of pottery in many places however, albeit at a more primitive level, whilst legends and religious cults clearly also survive. Nevertheless it seems fairly clear from this evidence that we are dealing with human disasters of almost unparalleled magnitude, and one is left wondering what natural or human agency could have influenced events on such a devastating scale.

The fascination and enduring interest in the Mycenean catas-trophe obviously lies not only in the fact of a dark age and in the fact

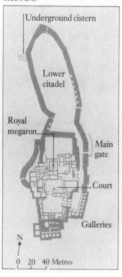

Plate 3(a). The actual layout of a Mycenean stronghold around 1250 BC, including the location of the royal megaron or small central shelter (from *In Search of the Trojan War* by M. Wood).

that it was preceded by widespread and apparently terminal destruction,[34] but in identifying what gave rise to the destruction. Furthermore, the interest which historians take in the subject is not purely for the sake of the Mycenean collapse itself. Throughout history, periods of splendour and strength have been superseded by depression and rapid collapse, many prosperous and powerful nations having been wiped away in course of time, for example China's warring states, the Mayan civilization and the empire of Rome; and yet in many of these cases, the causes of the decline remain equally obscure. As already remarked, it is not known whether we are dealing with some general rule relating to the internal evolution of society, which has utopias self-destructing, or whether some additional external factor is being overlooked. Any advance in understanding that relates to a particular downfall may therefore have an obvious bearing on the general phenomenon and hence the future of our own civilization. As one might expect, then, all sorts of possible reasons for the Mycenean collapse have been explored – crop failure and famine, widespread earthquakes, massive invasions, proletarian revolution, inter-baronial wars, to name a few – but although some of these may have played a part, there is now an increasing impression that something which transcends normal expectation may have been involved.

At first sight the problem seems more mundane: the Myceneans apparently recognized and anticipated the forces that led to their demise. It is a curious fact that, having replaced the Minoans after

Plate 3(b). Mycenean stronghold at Tiryns (photo: Peter Clayton).

1400 BC as the principal maritime-commercial presence in the Mediterranean, the feudal chiefs of the Mycenean nation during their period of dominance and supreme power became increasingly preoccupied with securing their defences at home. There was evidently some kind of awareness that the nation was coming under siege since look-outs were posted and defending forces were frequently mobilized in readiness for potential raiders. This activity has even been likened to that of the Viking period in Europe two thousand years later. Thus, although the Mycenean economy apparently flourished and elaborate palaces were built at Pylos, Mycenae, Tiryns, Iolkos, Gla, Orchomenos, Thebes and Athens, many of these edifices were also protected behind massive cyclopean walls. Attention has been drawn to the simultaneous revival of a peculiar ancient practice, namely the construction within the walls of the palace of a small covered enclosure known as a megaron, whose access was commonly through a low-level anteroom, ·curiously reminiscent of a modern air-raid shelter but of unknown purpose (plate 3(a)). Yet all this preparation was apparently to no avail. For, by the end of the twelfth century BC or thereabouts, all these citadels, apart from Athens, had been destroyed and abandoned. To what force did the classical Greeks attribute the destruction? There is little uncertainty here: the invaders were called Heraclids!

The context of the references to the Heraclids does not suggest that a figurative or metaphorical connotation was intended. Local rebellions or peasant uprisings following some breakdown in the economy or failure of crops are not inherently likely explanations therefore, even supposing the well-armed Mycenean aristocracy at the height of its powers, and its almost impregnable citadels, would lead one to suspect such an eventuality was reasonable. At the same time, despite the traditions, the study of archaeological remains from the end of the Mycenean age has failed to turn up any features which can be attributed with certainty to invading forces: burial habits for example and the weaponry employed reveal no significant changes. Nor are there any linguistic developments which might suggest the arrival of conquerors. Indeed the evidence, already commented on, points more to a continuity of development going back to the incursion of Indo-European speakers since 2000 BC. If the Heraclids were understood, it seems that they could only have been phantom-like raiders from an unknown northern territory who did not remain in any of the lands they overran.

In fact the whole of the eastern Mediterranean area (figure 1) was also overrun and devastated during this period. Not only did the Mycenean empire collapse, so also did its mighty Anatolian neighbour, the Hittite empire. In this case, the capital city Hattusas was engulfed in flames and destroyed as were the other Anatolian cities such as Troy, Miletus and Tarsus. To the east, so also were the main cities of the Levant, Alalakh, Carchemish, Qadesh, Qatna, Ugarit; and in Palestine, many other urban centres were likewise burned and ruined. So, quite apart from the events in the Aegean, the prosperity and stability of a far more extended area came to an abrupt end between 1230 and 1180 BC. The whole of this region sub-sequently descended into a period of almost total decline and comparative isolation whose intensity is truly hard to understand. In the Aegean the dark age lasted into the eighth century BC and in Anatolia almost as long. In the Levant the decline was less pronounced but real all the same, whilst in Canaan the old strongholds were fought over by the Philistines, Armaeans and Israelites. In the words of one authority, 'the evidence is easy to see in sacked cities, tumbled walls, broken communications, depopulation and deprivation'. Only Egypt, no less enfeebled for a while, managed to survive and maintain a semblance of civilization.

Here, there were merely the 'interregnum years', so called, some time prior to 1200 BC. A posthumous summary of events during the reign of the pharaoh Rameses III (1194–1162 BC) starts with an account of the preceding period in which it is clear that the breakdown was so complete that years went uncounted and records were not kept:

The land of Egypt was abandoned and every man was a law to himself. During many years there was no leader who could speak for others. Central government lapsed, small officials and headmen took over the whole land. Any man, great or small, might kill his neighbour. In the distress and vacuum that followed there came a Syrian, a foreigner who set himself up over the whole land, and men banded together to plunder one another. They treated the gods no better than men, and cut off the temple revenues.

Evidently we are dealing with an instance of total economic collapse and general disaster, leaving society hopelessly exposed to incursions from outside. But after the recovery had been made, Rameses III still had to ward off 'The Great Land and Sea Raids' between 1190 and 1180 BC.

All at once the lands were on the move, scattered in war. No country could stand before their arms. Hatti, Kode, Kizzuwaka, Carchemish, Arzawa and Alishaya. They were cut off. . . . They desolated its people and its land was like that which never came into being. They were advancing on Egypt while the flame was prepared before them.

Then later: 'As for those who came together on the sea, the full flame was in front of them at the river mouths.'

The Egyptian records therefore supplement and bear out the archaeological evidence in a remarkable way. They appear to distinguish three phases in the sequence of events. First there were the forces armed with a full flame who seem to have been an undoubted threat. It would not be surprising if these were the invaders who caused so much damage elsewhere. Then there were the unwelcome refugees from the devastated lands to the north. Only later was the pharaoh able to claim that he had succeeded in overcoming both of these problems; and by implication, the marauders who totally crushed the Mycenean and Hittite empires were held at bay by the Egyptians.

Yet, despite the convincing corroboration from the whole of the Mediterranean area, and the undoubtedly very real destruction, researchers have so far been quite unable to discover who these incendiary invaders might have been. The identity of the Heraclids, in modern times, has become a complete mystery.

4

The Sky Gods

WHATEVER one would like to imagine the Heraclids signify, there is little doubt as to what the Greeks themselves, emerging from the Dark Age, thought. They saw the Heraclids as sons of Heracles and the latter as a god.

At the same time, it also has to be admitted that the Greeks emerged from the Dark Age rather perplexed over the exact genealogy of the gods. Even as early as Homer (*c.* 800 BC) or Hesiod (*c.*700 BC) their heroic myths seem to have derived from a variety of sources.[35] Indeed different poets, and even the same poet in different passages, apparently preserve a number of very old traditions. The ones about Heracles are peculiarly complex; the elements absorbed into the tales of his exploits include conflicts with primeval monsters, ambivalence in his own nature since he is prone to fits of bestial frenzy, and his relationship with death and world-end. From the many adventures of Heracles, twelve came to be selected as the tasks imposed on him by Eurystheus, the king of Mycenae, and were depicted for example at Olympia around 560 BC.

It was said that Eurystheus, who was the son of Hera, Queen of Heaven, feared and envied Heracles, who was the son of Zeus, and hoped to destroy him by imposing on him twelve impossibly difficult and dangerous tasks. Heracles however successfully completed the 'twelve labours', one of them being against the city of Troy. Six of the tasks performed for Eurystheus are located in the Peloponnese and there is a suggestion that the stories were originally about a buccaneering feudal prince who may perhaps have been a real man. Such evidence indicates that the earliest Greek stories (prior to 700 BC) inherited elements going back in some form to the period five hundred years or so earlier when Mycenean warriors were in contention across the sea. It seems therefore that, by the fifth century BC, through the pervasive influence of the Olympian cult, notions had spread throughout Greece which intertwined the exploits of the gods with those of a few dimly perceived chieftains many centuries previously. For most Greeks – not all – the Mycenean period even came to mark the beginning of time itself, whilst Heracles became a symbol of strength and immortality, the greatest heroic figure of them all. He more than any other hero bridged the seemingly impassable gap between the short and fateful life of man and the unending splendour of the gods.

It was apparently through contacts further east that the early Greeks first learned to doubt their homegrown wisdom. The Greek historian Herodotus (484–430 BC)[36] for example, originally from Asia Minor, remarked that:

In Greece, the youngest of the gods are thought to be Heracles, Dionysos, and Pan; but in Egypt Pan is very ancient, and once one of the 'eight gods' who existed before the rest; Heracles is one of the 'twelve' who appeared later, and Dionysos one of the third order who were descended from the twelve. I have already mentioned the length of time which by the Egyptian reckoning elapsed between the coming of Heracles and the reign of Amasis; Pan is said to be still more ancient, and even Dionysos, the youngest of the three, appeared, they say, 15,000 years before Amasis. They claim to be quite certain of these dates, for they have always kept a careful written record of the passage of time. But from the birth of Dionysos, the son of Semele, daughter of Cadmus, to the present day is a period about 1600 years only; from Heracles, the son of Alcmene, about 900 years; from Pan the son of Penelope – he is supposed by the Greeks to be the son of Penelope and Hermes – not more than about 800 years, a shorter time than has elapsed since the Trojan War.

This is a rather confusing cocktail of divine appellations, but we can put aside Semele, Cadmus, Alcmene, Penelope and Hermes, all of them merely circumstantial details as far as our present purposes are concerned, and concentrate on Hermes, Dionysos and Pan. It is the recorded uncertainty over *their* relationships that is of immediate interest here, for it reveals to us how an informed traveller was seeking to improve the rather truncated view of world history that the Greeks of the fifth century BC still possessed.

Herodotus has been accused of magnifying the wonders and hoary antiquity of Egypt in order to adorn his own tale, and there have been suggestions too that his Egyptian informants were not without some degree of national chauvinism. However, the chronological sequence described by Herodotus is consistent in itself and does not contradict what we already know. Amasis, for example, was an

Egyptian pharaoh whose reign may be taken as a relatively secure marker, since it corresponds to the time, only a century before Herodotus, when Egypt and Greece were re-establishing regular contact after the Dark Age. There is evidence that the Egyptians at this time were considered by the Greeks to be counting in months rather than years, though it is possible also that the Greeks did not fully understand the Egyptian reckoning of time based on extant priest-lists going back to the first dynasty (around 3000 BC). Probably, therefore, the Egyptian timescale should be reduced by a factor of twelve or ten, although the exact figure remains unknown. Be that as it may, Herodotus goes on to make it clear that the popular Greek view relating to the arrival of the gods was in error and that Heracles was at least as old as 1800 BC. We should also take note of the fact that Herodotus sees nothing abstract or symbolic about the gods: he discusses them as real things, with specific genealogies and birth dates in a factual historical context.

The Greek view of the gods prior to the fifth century BC was indeed quite matter-of-fact. The gods were still invested with great power, moreover, and it was no less the custom in Greece than it had been in Sumeria and Egypt for feudal chiefs to claim gods as their ancestors and hence a divine right to rule. But among a people where it had never been possible for power to be held by a single ruling dynasty, such claims carried less conviction, and it seems to have been in Greece that the status of the gods came to be first seriously questioned. And from these doubts, there emerged a widespread view, stated by Euhemerus, for example, that the gods must have originally been real people who were later unscrupulously ennobled to supernatural status by their devious descendants. On this reckoning gods were a straightforward invention and the earlier matter-of-fact view a delusion. Although it could have been that the only invention was the investment of humans with divine rights, the idea that the gods were truly real also gradually became less attractive in the atheistic society that subsequently evolved in Greece. It hardly seems likely that ideas would have developed in this particular way if any visible basis for belief in real sky gods had remained in the sky. Ideas did so evolve however and there is no reason to be astonished by the matter-of-fact view of the gods espoused by the early Greeks. It seems likely that the very earliest Greeks regarded the Heraclids as perfectly real descendants of a perfectly real Heracles.

Heracles was also given an ancient ancestry in a religious cult ascribed to the probably mythical Orpheus.[37] Orphism was not at first widely accepted in Greece but the traditions of its theogony became well known eventually through the teachings of Pythagoras (c. 530 BC) and Plato (427–347 BC), having spread along with its associated mystery religion from Thrace into Greece. The stress in Orphic doctrine on the division between body and soul, the immortality of the latter and its possible reincarnation, suggest an earlier Egyptian connection. The Orphic school is therefore likely to have been quite ancient and there is some suggestion that it may even reach back to the fourteenth century BC, having been brought to the northern Aegean from Egypt by the fugitive priesthood of Akhenaton (1367–1350 BC).

Apart from its mystic rites, Orphism was noted for its speculations on the nature and relations of the various gods. One of the gods belonging to this sect was Aion, of whom it was said that he 'remains ever unchanged by virtue of his divine nature, who is one with the world, who likewise has neither beginning nor middle nor end, who partakes of no change, who created the whole of divine, living nature'. Aion, according to ancient authority, was born out of Water and Earth in the form of a snake with two heads, those of a bull and a lion. Between the heads was the face of a god with wings on its shoulders. The name of this strange deity was Cronos Ageraos (unageing time) or Heracles. In Greek tradition generally, Cronos was a remote primary god who had originally produced everything, divine or human, out of himself (plate 4).

Evidently then, there was some confusion over the actual provenance of Heracles. On the one hand there were those whose beliefs may have derived from the Egyptians and who saw him as quite ancient. On the other hand there also existed the view that the world only began around the time of the Trojan War, in which case Heracles was considered to be relatively young. Heracles remained an extremely significant deity on either account however, and we may assume that the Heraclids were well understood by the Greeks generally to be awe-inspiring figures, reaching back to the very source of creation itself. That this should be so is an important clue to the nature of the destructions.

The Orphic tradition also recounts how a god called Phaethon ran amok and set fire to the Earth. Homer in fact associates Phaethon with one of the labours of Heracles and provides a link with the events surrounding the downfall of Mycenae. The word Phaethon

Plate 4. Heracles. In late Roman (b) times, Heracles was considered to be identical to the ancient god of Time = Cronos = Aion, the latter being the equivalent of (a) Zervan, a winged god with a lion's head, in the Mithraic tradition; and (b) Phanes, a winged god born of a cosmic egg, in the Orphic tradition. Note that Zervan stands upon a world globe characterized by two crossed bands (cf. Figure 3), whilst Phanes stands amidst the twelve zodiacal signs, indicating that the egg has cosmic significance (see Van der Waerden, *Science Awakening II*). The topmost constellation in the zodiac here is the Ram = Aries (reprinted by permission of Kluwer, Academic Publishers. Photo: the Bodleian Library, Oxford).

actually means 'blazing star' and it may even have been another name for Zeus, the principal figure of the Greek pantheon. According to the Roman poet Ovid who wrote much later (around the time of Christ), the chariot of the Sun was taken over by his son Phaethon who then allowed it to wander from its course.[38] Phaethon failed to hold the reins and caused the horses to 'break loose' and 'rush aimlessly, knocking against the stars set deep in the sky and snatch . . . the chariot along through uncharted ways'. Then:

the earth burst into flame, the highest parts first, and split into deep cracks, and its moisture all dried up. The meadows are burned to white ashes; the trees are consumed, green leaves and all, and the ripe grain furnishes fuel for its own destruction . . . Great cities perish with their walls, and the vast conflagration reduces whole nations to ashes.

Apart from the description of processes which appear to have considerable relevance to events at the close of the Bronze Age, the reference to horses and ashes is doubly interesting because it is strangely reminiscent of that story of a wooden horse which, on some later accounts, specifically caused the downfall of Troy during the Mycenean decline. This particular course of events was never alluded to in Homer's *Iliad* and the story is generally seen as a post-Iliadic addition designed to reveal Odysseus as an unscrupulous strategist. It remains conceivable however that the story is cosmological symbolism with a basis in fact. The connection between wood and ash is obvious, as are the cosmic connotations of the Phaethon tale. The brightly flowing trails of fireballs, and the tails of great comets, have been associated since antiquity with the manes of horses.[39] Such visions in the dark sky are known to have always had terrifying associations, a distant reminder of which lingers on in our modern word 'nightmare'. In Arcadia, Poseidon was worshipped as the master of horses and was widely regarded as the instigator of earthquakes as well as being the brother of Zeus. Orphic theogony seems therefore to have an underlying naturalistic character and reveals potential links between events which may have actually happened and cosmic mythology. We shall be pursuing this link in more detail in due course.

This early naturalism – the linking of events on earth with the actions of the gods – is open to two interpretations, two models of reality. One can treat the stories with disbelief because, of course, supernatural intervention in the affairs of the world is absurd. Or one can take them seriously, the supernatural or divine connotations in this instance implying not the ghostly and the spiritual, but a disposition on the part of some real phenomenon to be larger than life and endowed with capacities beyond human control. It goes without saying that most commentators in modern times have selected the first alternative but if we now choose the second, on the grounds that something real must have been involved in the destruction of the Mycenean empire, then it will be necessary in due course to identify with certainty that period, later in history, when the meaning of divinity changed from the physically real to the merely figurative. For the moment, let us explore the connotation denoting realism a little further and see if it is possible to construct a bridge between the Egyptian and the Greek cosmos.[40]

Phaethon's relationship with the Sun is, in fact, a rather unusual one for the Greek gods. However, as we have noted, many aspects of the Orphic tradition are suspected survivals from the fourteenth century BC when they were first brought to the northern Aegean from Egypt by members of the priesthood of Akhenaton. The members of this priesthood adhered to the belief, radical in Egypt, that the Sun is the All-father. Thus, there is no need to be concerned about the solar aspect of the Phaethon story since it was generally foreign to the Egyptian and Greek conception of the Supreme Sky God, at least well into the first millennium BC. On the other hand, despite considerable differences, there are also aspects of the Greek and Egyptian traditions which show surprising agreement, especially those concerning the way in which the cosmos was made and how it subsequently evolved. These features in common might have arisen when contact was established between the two countries after the Dark Age, a possibility not generally given much credence by scientific historians;[41] or it might equally be due to a joint awareness of universal facts.

Both before and after Akhenaton, the Egyptians preserved a polytheistic tradition; there was however a leading figure amongst the gods. He was the ultimate source of existence, but He was definitely not the Sun. Originally this creative power was known under different names at the various cult centres in Ancient Egypt: Atum-Re at Heliopolis, Ptah at Memphis, Thoth at Hermopolis and Khnum at Elephantine. But He was essentially the same Being in all the cults and recognizably the agreed source of a number of specific great gods, Osiris, and Seth included, from whom there was believed to have sprung later a whole host of lesser sky gods as well as living things and mankind on earth. Indeed according to Memphitic theology, it was even said to have been through the thought and word of Ptah that the next generation of gods – the Ennead, which included the sun god Atum – had been brought forth from the primeval waters.

In these early Egyptian cosmological speculations, there was something that can easily be seen as having anticipated later Jewish and Christian doctrines. For example in the Fourth Gospel version of Genesis one reads that 'in the beginning was the Word, and the Word was with God, and the Word was God'. Furthermore, the emphasis on gods and men who were fundamentally similar in character and behaviour, as well as having a common origin, anticipated in considerable detail the cosmogony described much later by Plato. In Egyptian theology as in Greek therefore, there is little doubt that we deal with a cosmos in which the actions of gods and men in their respective domains were thought to be identical, and in which the relationship between gods and men was such that lives here on earth were part of some continuum that embraced life in the sky as well.

Now, in Egyptian cosmology, we have to draw a distinction between 'the world of the Horizon', namely our flat earth (Geb), and 'the world of the Two Lands' discussed earlier, which emerged from the primeval waters (Nun) and which seemed to incorporate the Sun's path. The latter, which is of course marked in the sky by the constellations of the zodiac, arched over and under Geb, the regions of the living (Shu) and dead (Duat) respectively, exposing sections known as Nut (heaven) and Nannet (counter-heaven). There is a tendency nowadays to think of heaven as somehow incorporating the whole of the visible sky but Nut was clearly envisaged by the Egyptians as a strip or path stretching across the sky in an east–west plane. It was said to be supported by four pillars from mountains at the edge of the Earth but there is no evidence that these pillars corresponded to the cardinal directions; rather it seems the path simply had dual characteristics, as if it divided into two interwoven strips. This dualism, associated with Ptah (the 'Lord of the Two Lands') and, as we have seen, Osiris and Seth, is likewise often linked nowadays with Upper and Lower Egypt but strictly speaking it relates to upper and lower reaches of the Nile. The Nile was, strangely, the *celestial* river of that name, commonly depicted in the vault of the sky as the goddess Nut, alternatively as an immense cow with a star-spangled belly or as a milky way traversed daily by the Sun and nightly by the personified stars or gods (Plate 5).

This distinction between the celestial and geographical Niles is one that both classical and modern commentators tend to be very casual over but it is one nevertheless that the ancient Egyptian worshipper was not likely to misunderstand. Thus:

thou makest the Nile in the lower world and bringest it whither thou wilt, in order to sustain mankind, even as thou hast made them ... thou hast put the Nile in the sky, so that it may come down for them and make waves upon the mountains like the sea, in order to moisten their fields and townships ... the Nile in the sky thou appointest for *foreign peoples* ... whereas the (other) Nile, it comes from the lower world for (the people) of Egypt.[42]

The 'world of the Horizon', namely our flat earth, and the 'world of the Two Lands' were also not infrequently at odds with one another. The state of great tension that was considered to exist in Egypt was thus expressed in terms of the struggle between Horus and Seth on the one hand and the pharaoh on the other, the latter clearly posing as a mediator between the unified nation on earth and the divine forces responsible for cosmic order. The point has already been made that the divine forces were very real and even brutal so far as the Egyptians were concerned since it was evidently not taken for granted that the world was safe. To counter the dangers, the pharaoh essentially became the stabilizing hub of the nation, uniting it with the divine influences and controlling agencies of the universe, and virtually turning into a cosmic figure himself. One can only assume that the pharaoh, as a mere mortal, knew what he was about and was able to secure unto himself this degree of omnipotence through demonstrating a real ability to anticipate and avoid the recognizably malevolent cosmic forces, whatever they were, either through chance or skill.

Modern interpreters, of course, have tended to assume that these malevolent forces were fictional and they therefore picture the Egyptians as a rather gullible people capable of swallowing every aspect of a demonstrably false and purely invented cosmology. However, it is equally plausible, if not more so, to suppose that these people were much like ourselves and that they may have been caught up in events that demanded urgent solution and that, as usual, they

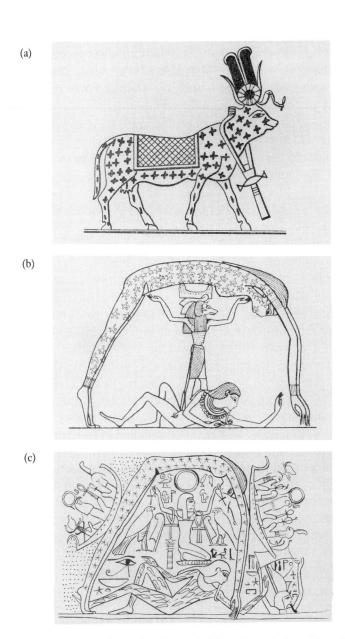

Plate 5. In the writings of early Greek natural philosophers, the heavenly path of the Sun was associated with the Milky Way. A similar personification of the Sun's path is found in ancient Egypt where it is represented by (a) the star-spangled cow-goddess Hathor, later Isis (line drawing reproduced by permission of Weidenfeld & Nicolson), or by the similarly decorated mother-goddess Nut, not bearing (b) or bearing (c) the solar barque as the case may be (line drawings reproduced by permission of Crossroad/Continuum).

largely deferred to experts who came up with what appeared at the time to be the most rational explanation of what was going on. If so, it is understandable that the pharaoh should have employed a professional priesthood and conferred upon it considerable privileges and power as well as astronomical duties. And if the cosmos forever remained beyond their capacity to truly comprehend, then it is understandable that cosmological theory should have developed or evolved sometimes perhaps by desperate edict from the pharaoh.

In the New Kingdom for example, the city of Thebes rose to pre-eminence as the principal sacred city having absorbed various cosmogonies (Memphite, Heliopolitan and Hermopolitan) and their respective systems of gods. Its priesthood advanced the worship of Amon-Re, the body and face respectively of Ptah who controlled and pervaded the whole universe, together with the heavens and the underworld. By this time, the concept of a single universal god was making itself felt more and more, and no great principle would seem to have been at stake when Akhenaton decided that Atum, the monotheistic power manifested in the disc of the Sun, should be exalted as the one and only god. Nevertheless, Tutankhamun restored the succession under Amon-Re after the death of Akhenaton and he was said to have 'driven out the disorder from the Two Lands, so that order was again established in its place, as at the creation'. It seems therefore that just around the time the Minoan civilization was undergoing its decline, radical changes were being introduced by a new priesthood under Akhenaton which relate to priorities in the sky. It is by the way that an error was later thought by the Egyptians to have been committed at this time, though it is evidently important to know that those responsible for the change incurred the wrath of the people and subsequently escaped with their ideas to the Northern Aegean where they seem eventually to have re-emerged as part of the Orphic doctrine, in which the divine cocktail evidently embraced some surprisingly naturalistic elements.

When we turn to early Greek cosmology, we find its various facets closely resemble those of Egyptian cosmology. Thus the earth (Ge rather than Geb) now appears as a vast plain encircled by a stream (Ocean) resembling the waters of Nun in Egypt, from which the gods and mankind had earlier sprung. Likewise, Heaven and Earth were thought to meet at some points on the extremities of the horizon, at the so-called pillars of Heracles, the entrance to the underworld being situated in the west where the Sun descended into the nether regions whilst the stars arose on the other side. There was also some idea that Hades, the counterpart of the Egyptian Duat, could be reached through rifts in the rocks at certain places on the earth's surface. The Greeks, like the Egyptians, did not at first consider the universe and mankind to be transcended by the gods. On the contrary, the gods shared human weaknesses, passions and intrigues, and they were generally attributed no higher status than conquering chieftains.

Somewhere in the background though was the rather remote figure of Cronos who could be likened to the Egyptian Ptah. The nineteenth-century view that Cronos – the Roman Saturn – was identical with Time (Chronos) has been considered somewhat questionable during the present century; however if Cronos was in fact a recognizable body in an Earth-crossing orbit of short period, there is no reason, in principle, why its regular close encounters with the Earth should not in the past have become the basis of time-keeping in the form of a sacred calendar, with successive years beginning in November, say, in addition to the one we now associate with the solar year. Sacred calendars of unknown origin are evident elsewhere in the ancient world (e.g. Mayan and Indian calendars). On the face of it, then, there is no compelling reason why Cronos and Time should not in fact derive from a common source in the very remote past (see also plate 4).

Be that as it may, it is the creative role of Cronos that we emphasize here. As with his Egyptian counterpart, a kind of dualism is apparent. Cronos eventually fathered two bands of gods, the horrible Cyclopes and the rock-hurling Titans, who engaged one another in battles, leaving the Earth sometimes as the unhappy but innocent victim. It may be that the mighty Zeus and the very considerable Poseidon (or Atlas) were the principal contestants here but in due course, the former alone seems to have inherited the cosmic arena. Zeus, who was certainly of the same generation as the Titans, and who is now thought to be identical with the second-millennium Babylonian god Marduk (see chapter 1), was also the main Indo-European sky god and weather god, combining the functions of a nature god with those of the head of an anthropomorphic pantheon mostly fathered by himself. From his exalted abode in Mount Olympus, probably in the sky, this divine 'cloud gatherer' poured refreshing fertilizing rain on earth, but he also blasted gods and men alike with his terrorizing thunderbolts when they thwarted his irascible will or frustrated his plans and purposes. Eventually however, even Zeus became a less anthropomorphic figure and turned more into a divine cosmic presence exercising his will through more subtle and invisible means.

It is a difficulty confronting the mythologist that ideas about the nature of ancient gods did gradually evolve. Nevertheless it is possible to construct a general correspondence between the Mesopotamian, Egyptian and Greek theogonies, thereby confirming their basis in a common source or sequence of events. In the later versions for example we are aware of variously named remote father-figures who eventually came to lose some of their pre-eminence: Anu in Mesopotamia, Ptah in Egypt and Cronos in Greece. For both sets of progeny fathered by the latter figures, a kind of dualism involving an element of conflict was clearly significant, as were the new dominant figures that emerged, Horus in Egypt, for example, and Zeus in Greece. By the mid-second millennium BC however, the supposed backdrop of events justifying the pre-eminence of these gods was apparently on the wane, and the characters of Horus and Zeus seemed to evolve further. Just as Zeus became less anthropomorphic, so the guise of Horus adopted by the pharaohs became the guise of Amon-Re, also a divine presence who apparently pervaded the whole universe. The remoteness and pervasiveness of these figures in due course seemed to supplant that of Ptah and Cronos whilst the more immediate celestial stage of the pre-classical period may have been taken over by a powerful new god whom both the Greeks and the Egyptians called Heracles.

In the principal wars of the gods as described by Egyptian and Homeric sources, Heracles, as we have seen, was the offspring of Zeus or his equivalent. The Orphic variation, if it did not arise from an overlapping of stories, suggests that Heracles corresponded to a reappearance in a new world of a very ancient god. Recurrence indeed became a very important theme in Greek cosmogony but whether it originated with the Orphic tradition, or is much earlier, is a question for the historians and the mythologists. In general, it is clear enough that the gods had offspring and that there was a self-evident association between Heracles and Heraclids. Moreover, it is evident that through the Homeric association of the Heraclids with the downfall of the Mycenean nation, the god Heracles had every reason later to become a more significant figure in the Greek pantheon than he might otherwise have become. Thus the inherent contradictions and uncertainties that unquestionably exist need not be seen as a great difficulty for analysts but more the result of a known evolution in ideas about the nature of gods in the pre-Christian era.

To put the developments in Egypt and Greece in perspective,[43] it is now fairly clear that the gods and their creator were seen for the first time during the third millennium BC as perfectly real bodies in the sky who were primarily concerned in sorting things out for themselves and only indirectly concerned with the fate of men on earth. An important part of the creator's work at this time was an apparently visible world in the sky which lay close to the Sun's path and which embraced the Earth, a vision probably no less awesome than the gods themselves and a perpetual source of concern. Man's terrified attempts at appeasement, even to the extent of human sacrifice in some quarters, were a somewhat desperate measure whose efficacy remained uncertain. By the second millennium BC however, the conception of things had evolved into a formal dualism in which one particular great god, father of many others, was seen to be all-powerful and basically benevolent, whilst another particular god, also the source of many others, was no less strong but essentially evil. Both sets of gods, the angels and demons of subsequent myth, seem to have been looked on as members of a visible milky way which evidently corresponded to the path of the Sun rather than the Galactic system to which we now append the name. These gods had characteristics which we now recognize as cometary. Their history seems to imply the disintegration, over several millennia, of a very large comet, the debris producing bands of light in the sky and occasional devastation on the Earth. An orbit taking the comet and its offspring along the constellations of the zodiac is implicit in such features as the enclosure of Blackland, the flight path of Phaethon, the association of the Orphic Phanes or Heracles with zodiac (plate 4), and the otherwise inexplicable classical references to the zodiac as a 'milky way'.

Subsequently, during the first millennium BC, the malevolent gods appear to have departed, leaving a single benevolent god who became identified as both creator and director of a recurring universe whilst at the same time becoming himself increasingly remote and less tangible. The precise timing of these developments may not be clear, but the gradual advance towards a new brand of monotheism based on an invisible god who was also benevolent seems eventually to have given rise to a controversial technical problem, namely the true source of evil in the universe. The problem was eventually resolved by removing evil from the sky altogether, where it had always been, and treating it from now on as a fundamental aspect of the human condition itself.

There need not be too much doubt then that the Greeks in general, and the Athenians in particular, knew what the Heraclids were and whence they had come: they were evil bodies and they came out of the zodiac. But in their subsequent theological conflicts, nullifying the concept of sky-borne evil, the Greeks seem to have convinced themselves that the bodies did not exist. Let us now consider how this came about.

5

Renaissance

SOLON seems to have been a man of exceptional ability. He is first heard of during a critical period of the war between Athens and Megara at the end of the seventh century BC when he persuaded his fellow citizens to persevere in what seemed a hopeless contest. Then, having saved the city from ruin and bloody revolution, Solon was elected an archon in 594 BC and entrusted with the duty of drafting a new constitution.

Solon appears to have been both just and liberal whilst giving priority to restoring confidence amongst the manufacturing and trading industries: thus he forbade the loan of money on the security of the borrower's person and he cancelled all outstanding debts. He revalued the coinage so as to put Athenian traders on terms of equality with their Ionic counterparts. He also abolished all privileges of birth and instituted a public assembly, or Ecclesia, upon which he conferred considerable power. It was a function of this assembly to debate affairs of state and to establish through democratic vote the natural, moral and public laws which were then to be promulgated as decrees by the archons. Displaying unusually profound wisdom and considerable practical ability, Solon seems to have proved himself a remarkable legislator and it is a reflection of his considerable stature and the veneration in which his memory was held that his laws were duly inscribed on wooden pyramids which stood on the Acropolis for many years. Indeed, they were somehow preserved even until the time of Plutarch around AD 120.

In later years, Solon appears to have adopted the role of elder statesman or ambassador at large for Athens. It is even said that he exiled himself on purpose for ten years in order to avoid the importunities of those who were urging him to supplement his legislation with further clauses. Be that as it may, Solon travelled widely, going as far as the limits of the old Mycenean empire, to Asia Minor, to Cyprus and to Egypt, for example. We know from

Homer's account that some of the invaders who had attacked at Troy also found their way to Egypt and it may well be that a tenuous link with the centre of the empire had been maintained after its downfall. The account may of course refer to a link between Heraclids and Sea-Peoples as much as between humans. Either way, it is known[44] that:

in the Egyptian delta . . . there is a certain district which is called the district of Sais and the great city of the district is also called Sais . . . the citizens have a deity for their foundress, she is called in the Egyptian tongue Neith and is asserted by them to be the same whom the Hellenes call (Pallas) Athenae: they are great lovers of the Athenians, *and say that they are in some way related to them*.

It can be supposed therefore that Solon had ample opportunity through contacts with fellow countrymen during his visit to Sais around 560 BC, to acquire an accurate picture of Egyptian knowledge. There seem therefore to be rather good grounds for supposing that a wise old priest of Sais whom Solon consulted would have been a reliable informant and that what he reported would have been scrutinized with the utmost care by the wise old mandarin from Athens. It is at least possible therefore that we are dealing with remembered fact and that we should take seriously what Solon reported back to his fellow-Athenians on his return:

O Solon, you [Greeks] are all young in your minds which hold no store of old belief based on long tradition, no knowledge hoary with age. The reason is this. There have been, and will be hereafter, many and divers destructions of mankind, the greatest by fire and water, though other lesser ones are due to countless other causes. Thus the story current also in your part of the world, that Phaethon, child of the Sun, once harnessed his father's chariot but could not guide it on his father's course and so burnt up everything on the face of the earth and was himself consumed by the thunderbolt – this legend has the air of a fable; *but the truth behind it is a deviation of the bodies that revolve in heaven around the earth and a destruction, occurring at long intervals, of things on the earth by a great conflagration*. . . . Any great or noble achievement or otherwise exceptional event that has come to pass, either in your parts or here or in any place of which we have tidings, has been written down for ages past in records that are preserved in our temples; whereas with you and other peoples again and again, life [had only just] been enriched with letters and all the other necessities of civilization *when once more, after the usual period of years, the torrents from heaven [swept] down like a pestilence, leaving only the rude and unlettered among you*. And so you start again like children, knowing nothing of what existed in ancient times here or in your own country. . . . To begin with, your people remember only one deluge, though there were many earlier; and moreover you do not know that the noblest and bravest race in the world once lived in your country. From a small remnant of their seed you and all your fellow citizens are derived; but you know nothing of it because the survivors for many generations died leaving no word in writing.

This extract is taken from Plato's dialogues *Timaeus* and *Critias* in which appears the famous story of Atlantis. The Egyptian dates for the founding of Athens and Sais, which were said to precede the famous event, are respectively given as 9000 and 8000 years ago which, by the typical reckonings of the time, as noted for example by Herodotus or Eudoxus, imply 1450/1350 BC or 1300/1200 BC approximately. Either of these is coincident with the period of the Mycenean dominion. Moreover, as Critias reports '9000 years have elapsed since the declaration of war between those who lived outside and all those who lived inside the Pillars of Heracles'. Pillars, as we have noted, emerged from a cosmic enclosure which separated Blackland, including the Earth, from Redland, once the domain of the destructive Seth. Thus the drowning of Atlantis seems to relate to events observed in the sky at or before 1200 BC, the apparent immersion of an Island of Creation in an extended zodiacal sea, events during which mankind also suffered at the hands of invading hordes from the zodiacal band. The Atlantis story deals then with a seemingly visible assault from the cosmos whose structure was broadly known, wherein there were gods from whom it was customary to claim descent, and in whom it was customary to invest overlordship and ultimate power. Critias was thus able to carry the story back to an even earlier period in Greece and state how:

Once upon a time the gods divided up the Earth between them – not in the course of a quarrel; for it would be quite wrong to think that the gods do not know what is appropriate to them, or that, knowing it, they would want to annex what property belongs to others. Each gladly received his just allocation, and settled his territories; and having done so they proceeded to look after us, their creatures and children, as shepherds look after their flocks. They did not use physical means of control, like shepherds who direct their flocks with blows, but brought their influence to bear on the creature's most sensitive part, using persuasion as a steersman uses the helm to direct the mind as they saw fit and so guide the whole mortal creature.[45] The various gods, then, administered the various regions which had been allotted to them. But Hephaestos and Athene, who shared as brother and sister a common character, and pursued the same ends in their love of knowledge and skill, were allotted this land of ours as their joint sphere and as a suitable and natural home for excellence and wisdom. They produced a native race of good men and gave them suitable political arrangements. Their names have been preserved but what they did has been forgotten because of the destruction of their successors and the long lapse of time. For as we said before, the survivors of this destruction were an unlettered mountain race who had just heard the names of the rulers of the land but knew little of their achievements. They were glad enough to give their names to their own children, but they knew nothing of the virtues and institutions of their predecessors, except for a few hazy reports; for many generations they and their children were short of bare necessities, and their minds and thoughts were occupied with providing for them, to the neglect of earlier history and tradition. For an interest in the past and historical research come only when communities have leisure and when men are already provided with the necessities of life. That is how the names but not the achievements of these early generations came to be preserved.

Notwithstanding the lacunae in Greek history for reasons which Critias recounts, we appear to be the recipients, through Plato, of an old, reasonably coherent, Egyptian description of the role played by the ancient gods both in the sky and on the ground. The underlying theme portrayed by Plato is that of a mighty empire subtly controlled from the centre, a concept which at the time would have departed from the cut and thrust of Greek feudalism. Plato's main interest was political analysis and much of his thinking did eventually provide some of the inspiration behind the growth of the future Hellenic empire. Fortunately for our present purpose, however, Plato was also concerned to understand the astronomical phenomena involved and he records how Socrates persuaded Timaeus to give a description of what went on. Thus God began with a spherical body apparently containing the elements of soul and matter; out of this creation, he wove two circular bands,[46] similarly created but inclined to each other, in which were created the various gods – not excluding the Sun, Moon and planets – who were responsible for producing the living things on Earth and imparting soul to them. Despite frequent scholarly attempts to identify these two bands with more modern imaginary circles on the sky (figure 3), namely the celestial equator and the ecliptic, the account by Timaeus is quite definitely of observed bands and real gods: it seems much more likely that the dual structure of the former cosmic Nile is in fact

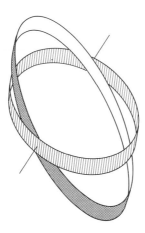

Figure 3. In the Timaeus, the 'body of the heaven' is said to be visible but it is also described as a strip of fabric cut down the middle to form two bands which are bent into relatively inclined circles fastened at their cross-over points, the celestial motions associated with the inner circle being specifically described as different, as would be expected for a body in an eccentric orbit.

being described. One of these bands is said to include the Sun, Moon and planets, but the latter had only recently begun to impinge on the Greek scientific consciousness. Timaeus in fact indicates that he has other bodies also in mind and if we are to consider his account in any way relevant to other aspects of the Atlantis story, it seems that these other divine bodies, rather than planets, must be the ones to associate with the 'torrents from heaven' that sweep down 'like a pestilence . . . after the usual period of years'. Thus, he says that 'they are so bewildering in numbers and so amazing in intricacy' that many are 'virtually unaware that their wandering motions are time at all . . . it is beyond our power to know or tell about the birth of these other gods . . . both those whose circuits we see in the sky and those who only appear to us when they wish'.

We have therefore from Solon an explicit statement that the Mycenean collapse was caused by an encounter with a swarm of astronomical bodies, giving rise to a great conflagration. The surprise to Solon of course was not so much the facts, for the Phaethon story was already well known, but the remarkably naturalistic terms used by the old priest at Sais to describe events which were commonly regarded as the work of gods. The advocate of open government back home thus learned that underlying the panoply of religion and popular magic was an understanding of the world that needed no recourse to divine agency: instead, man was just an innocent victim of recurring encounters between freely moving bodies in space!

To what extent this particular contact catalysed contemporary Greek thinking must be uncertain; but the fact is that the first Greek attempts to understand the universe in naturalistic terms, rather than as something controlled by the caprice of the gods, date from this time. These attempts began with a group of natural philosophers from Ionia, shortly after 600 BC. Ionia, a coastal strip of land which is part of Asia Minor, was geographically an interface between the organized, conservative Mesopotamian societies and the less centralized, more volatile Greek world. It was therefore a region in which thinking Greeks would compare their own traditions with those arriving from distant lands. However mere comparison of gods and the like would not on its own have triggered the origin of natural philosophy in Ionian Greece. Both Egyptian and Mesopotamian principles had to be put into the melting pot. From this time the Greeks, reared on a diet of Homer and still bearing the cross of the Mycenean tragedy, determined to explore the material world and expose it to the light of day. In so doing they began a process that eventually gave rise to western civilization.

In these days when good and evil are frequently regarded as terms to describe the moral choices made by individuals, it is difficult to conceive of a climate of opinion in which they were absolute entities ordained by a cosmic Deity. Yet if one is to appreciate the enormity of the change that now overtook ancient Greece during the sixth and seventh centuries BC, it is necessary to realize how deeply the old viewpoint was impressed at this time upon the peoples of western Asia, the Fertile Crescent and the eastern Mediterranean. The ancient Egyptians in particular were outstanding in the interest they displayed in the moral aspects of the cosmos. Thus in their complex conception of the world, as we have already seen, the gods, the universe and the sovereign ruler were inseparably combined in a fundamental enduring principle: 'the good and worth of which was to be lasting. It had not been disturbed since the day of creation [and] he who transgresse[d] its ordinances [was] punished'. Seth, as the essence of evil, with the characteristics of the Devil, was certainly opposed to Horus but collectively the gods were regarded as good and the guardians of morality even if some might have malevolent qualities themselves. Visitations of disaster on earth were thus understood quite clearly as the acts of orderly cosmic gods upholding the right and true. The peculiarity of the Egyptian ethos however, which might not have been anticipated, was the extension of this principle to everyday behaviour for the purpose of maintaining an orderly society. Thus, according to the Egyptians, it was the height of wisdom to be just, righteous, godly and humane; and it was supposed that the ethical principles involved here, coupled with a sense of social justice and truth, were provided individually through some kind of divine consciousness. One was brought up accordingly to accept that the gods were righteous judges having knowledge of everything that took place and watching the actions of every man, punishing wrongdoing and rewarding virtuous acts. Such was the level of moral ascendancy implicit in this understanding of the cosmos and the state that it was generally felt to be useless to try to alter that which had been predetermined by the gods or to pry into their intentions and decrees. Yet, underlying this superstructure of state-imposed belief, as Solon was made aware, was a detached materialistic view of the world without any moralistic connotations at all.

In Mesopotamia likewise, and diffusing throughout the Near East, the same fear of heavenly retribution dominated beliefs; but here, there was less of a feeling that the gods were fundamentally good. As in Egypt, every phase of existence – cosmic, civic, domestic and personal – was brought within the orbit of the gods; that is, ethics and religion were inseparable, but the emphasis in this case fell more clearly on offences that were committed unwittingly. The public attitude to the gods therefore was apparently far more ritualistic, far more given to the belief that disaster would come regardless of any human penitence. In view of such fatalism, there was less concern on the part of the state regarding fundamental or private beliefs – benevolent liberalism perhaps rather than benevolent paternalism. Unlike the pharaohs therefore, the Mesopotamian kings were not cosmic figures though they were still very concerned personally to discharge their duties in the eyes of the gods by maintaining the appearances of state order. Oaths may have been taken at the temple or its gate, and referred to the judgement of God, but the judges were not priests. The Babylonian code and the administration of its legislation, even as far back as Hammurabi (about 1750 BC), were essentially secular and made no claim to have been divine pro-nouncements, in spite of their having been issued with the sanction of Marduk (Zeus). The responsibility for the well-being of the state seems therefore to have rested more directly on the king and he could thus represent himself as suffering vicariously on behalf of his people, a fundamental notion that persisted through into the Christian era.

The loose federation of Greek nation states never became a nation in the modern sense, and Athens only achieved its dominance, becoming a centre for the diffusion of culture, after the fifth century BC. But in the actions taken by Solon, we see the early signs of an inventive community advancing towards a form of government that differed radically from that of the Egyptians, Mesopotamians and other Near Eastern nations. Although there is no evidence of any conscious attempt to turn away from pharaohs and kings, Solon seems to have played the role of a head of state who wished to be both a detached and disinterested figure. Civic affairs and the nature of the material world, it seems to have been decided, were matters now to be settled by rational and public argument. The responsibil-ity for the well-being of the state was to be shared equally, in such a manner that the conscience of the first among equals was free. It was a novel development, with the good fortune to be supervised over two and a half centuries by men of genius like Solon, Pisistratus, Cleisthenes and Pericles, but would it work? Indeed, would the disappearance of the gods from the sky, real or imagined, affect the result?

6

Enlightenment

So long as the gods were real, visible and powerful entities in the sky, mankind's understanding of nature was likely to be in terms of their complex comings and goings, their genealogies and fadings, and their potential for capricious and devastating intervention in the affairs of Earth. The corollary is that tales of the kind brought back by Solon, in which the devastation involved not the arbitrary actions of gods but the wholly natural action of bodies from space, would not have been credible if these gods were still to be seen. The rise of *theory* in the Greece of the seventh century BC, in which the cosmos was explained in natural terms rather than through divine wilful action, could only have taken place if the sky gods themselves were fading or had disappeared from sight. There was a struggle, of course: Anaxagoras was exiled for casting the Sun as a red-hot stone, whilst Socrates was executed for espousing invisible gods. But eventually the belief in a single non-interventionist god was established, together with the notion that nature was set to run on its predestined course with only an occasional nudge from on high. Again, the rise of this movement could only plausibly have taken place against the backdrop of fading, impotent deities.[47]

The transition from myth to logic, from story to theory, was never complete and did not take place all at once; indeed it is likely that patterns of myth affected the patterns of rational thought with which the early natural philosophers emerged. Thus what appear to be real celestial events are described as a story, the end of Atlantis, by Plato as late as the fourth century BC; and not unexpectedly, the earliest theories of nature contain elements appropriate to a night sky which was different from the one we see now.

Thales (625–545 BC) was traditionally the first of the natural philosophers, but Anaximander of Miletos (610–547 BC) who followed him is the first whose theorizing is preserved in any detail. A recent reconstruction of Thales' cosmogony runs along the following lines: first there was a great ocean in which a circular eddy developed and the Earth grew. Round about the air was driven by the whirling currents whilst luminaries apparently soared upwards with the wind and then plunged down again through the water. The whole of this cosmos, which could run down and vanish only to start up again later, was borne on the sea like a vessel, controlled and steered by the water itself which, immortal and ageless, mover of things outside it, shaker of the Earth, also has the attributes of divinity.

Anaximander, likewise, seems to have envisaged a 'boundless' medium but in his case things started with a fiery egg-like formation out of the boundless, from which subsequently grew or separated off various rings enclosing one another like the barks of a tree. These rings, described as 'temporary gods', appear to have surrounded the earth which was cylindrical in shape, being three times as broad as it was deep, the upper face being where we lived. The whole of this system then hung freely in space, remaining in position by reason of its equal distance from everything else. The rings were clearly understood as coming beneath the Moon and were like hoops of compressed air, enclosing fire inside them, but with vents from which the light of stars beamed forth.

It is difficult to see how such apparently bizarre ideas could have arisen unless they had a naturalistic basis. The notion of stars as wheels filled with fire, with openings through which starlight is emitted, is particularly strange to modern eyes. That these 'stars' were seen as the nearest of the heavenly bodies, closer than Sun or Moon, is also strange because occultations of normal stars, their sudden disappearance when the disc of the Moon passes in front of them, are not uncommon and must have been known to Anaximander. It may be then, that in these very early cosmological speculations we are in fact detecting elements of a night sky *which is not the one we see now*, and that here we have an indication as to the nature of conspicuous bodies in the pre-classical sky which, like the ancient gods, were subsequently to disappear.

The Pythagoreans at this time were already of the opinion that the Earth itself was a star (presumably a planet) and that it moved around a central fire designated as the Mother of Gods or the Citadel of Zeus. It is possible therefore that the underlying picture was of the planet Earth in motion around a central source of fire, and of this fire displaying itself as stars shooting through a window in the sky whenever the Earth crossed hoops of fire perceived as an extension of the central source. Contrary to the usual belief nowadays, the orbital motion of the Earth was probably well understood by the early Greeks and was never forgotten. We certainly know from Cicero much later that Nicetus of Syracuse considered all the stars to be at rest and the Earth alone to be in motion; the same philosopher, he added, had shown that the Earth, by a movement of its axis, produced the same appearance in the heavens as if these were in motion and the Earth at rest. Aristarchus of Samos (310–230 BC) likewise anticipated Copernicus in almost every respect. Indeed, it is likely that this viewpoint was already present in the time of Pythagoras for he too conceived that a movement of the Earth about its centre caused the succession of night and day.

The question is then whether identification can be made of a central source in the ancient sky, containing within itself hoops of fire. There is no evidence that the Pythagoreans looked upon the central fire as the Sun, and the only plausible alternative, if they had a real object in mind, is the zodiacal cloud. Often overlooked by modern scholars, the zodiacal cloud is a discus-shaped swathe of comet dust through which the planets constantly move and whose most concentrated region is visible in good climates close to the Sun after dusk and before dawn. At the present time it can be seen as a pillar of light no less bright than the Milky Way, the pillar appearing close to the vertical in equatorial latitudes. If the zodiacal light were at one time much brighter, and contained decaying comets within it, then it would be seen to contain structure, and the Earth in the course of its orbit would run through these structures, or 'hoops of fire', sometimes to spectacular effect. In this connection it is of interest to mention a conjecture attributed to Democritus of Abdera, who lived about 450 years before Christ, and whom Seneca designates as the most subtle of all the ancients. According to Seneca, this writer held that there were many bodies circulating about the centre of the universe but invisible to us on account of the obscurity of their light or their positions in their orbits.

It is not unlikely that Pythagoras gained much of his knowledge from the Babylonian school (chapter 1); and although the initial conception of an Earth-centred world is now popularly associated with the very ancient civilizations, Seleucus, a distinguished Greek of the Babylonian school a century after Aristarchus, was still maintaining quite correctly that the Earth not only had a diurnal

Plate 6. Visible debris from Encke's Comet or its progenitor: the Zodiacal Light from South Africa, 7 June 1843, as drawn by the Edinburgh astronomer Charles Piazzi Smyth. The light is aligned along the path of the zodiacal constellations, and was the track along which, in remote antiquity, active comets from the Encke material moved (by kind permission of Don Africana Library, Durban, South Africa).

rotation but also an orbital motion around the Sun. We cannot therefore exclude the possibility that the Babylonians of this period always had a reasonably accurate impression of the underlying structure of the Solar System which somehow became distorted in the hands of the Greeks. It could not have been an impression that was ever in the forefront of Babylonian thinking however, for if our understanding of the pre-Socratic philosophers is correct, the Babylonians and others of the period would have been preoccupied with meteors and fireballs and their association with comets. Indeed, if they understood meteors and fireballs correctly, as the products of dying comets in hoop-like orbits that were no longer visible, there would have been a quite natural awareness of 'hidden stars' coming between us and the Moon. Something of this knowledge may have been preserved in the teachings of Anaxagoras (c.500–428 BC) for he held that the Earth is flat but:

that the Sun and Moon and all the stars are fiery stones carried round by the rotation of the aether. Below the stars are the Sun and Moon and also certain bodies which revolve with them but are invisible to us. . . . The Moon is eclipsed by the Earth screening the Sun's light from it and sometimes, too, by the [other, invisible] bodies below the Moon coming in front of it.

In fact, the pre-Socratic philosophers generally, drawing upon Egyptian, Babylonian, Persian and other ancient sources, appear to describe a world that came into being through the production and concentration of four elements (fire, air, water, earth) from the original undifferentiated mass or cosmic egg. The world formed of these elements was basically a temporary flat system in the zodiac. It included the flat Earth surrounded by the Ocean. The Ocean surrounding the Earth was supposed to be continuous with the sky or 'waters above'. A succession of such worlds was apparently envisaged, each manifestation being in due course destroyed and returned to the so-called 'boundless'.

It is remarkable that this is exactly the picture we expect for an extended series of enhancements of the zodiacal light (plate 6) by a disintegrating comet. The disintegration of a major comet or its offshoots injects a large mass of dust into the zodiacal cloud, and since the latter is seen by reflected sunlight, the result is a temporary increase in its observed brightness. Rare giant comets, orbiting in the inner planetary system with periods of only a few years, are a likely major source of such debris although no such active comet is visible at the present time (see chapter 10). These early accounts may nevertheless be understandable as attempts to describe the process by which an active comet of this kind periodically enhanced the zodiacal light at some past epoch, creating clouds of debris which slowly spread along the constellations. In fact, there are several indications that the earliest references to the Milky Way are descriptions of an intense zodiacal light and that it, too, was able to reach below the Moon. It was held, for example, that the Milky Way was formerly the path of the Sun or that it lay in the Earth's shadow produced by the Sun. Furthermore, it was a hot accumulation of the disintegration products of many comets.[48]

Jets issuing from rotating wheels of fire; bodies coming between Moon and Earth; temporary 'worlds' forming in the plane of the zodiac: it seems reasonable to conclude that the earliest philosophers were describing, perhaps from the experience of their forebears, an essentially correct association between cometary disintegration products and the formation of a luminous dust cloud in the plane of the ecliptic, albeit one which was also supposed to come between us and the Moon. We are beginning to see, perhaps, hints of a night sky which was not the one we see now; and perhaps even clues to the nature of the Heraclids. The gods and their thunderbolts, now on the wane, were merely comets and fireballs entering a less active phase.

Until recently, it has been assumed that statements about the cosmos provided by the pre-Socratic natural philosophers are best understood in terms of planetary rather than cometary phenomena, and it is an important question therefore whether cometary phenomena were directly observed or whether the knowledge of such phenomena was derived from earlier historical sources no longer extant. Obviously the Pythagorean link might have been important here, providing an input of knowledge from very early Babylonian sources. On the other hand, if the destructions of the Minoan and Mycenean civilizations (chapter 3) bear witness to a then fairly recent but substantial comet disintegration in the sky, it is to be expected that some fairly major debris would have enhanced the putative Milky Way in the zodiac during the Dark Age, that is, during the previous several hundred years. As we know, these phenomena would then be no earlier than the exploits of Heracles and the Heraclids: historical sources relating to these phenomena would then in principle be no less accessible than the sources available to Homer and Hesiod.

If the sky was in fact different in the first millennium BC and before, this might also account for the seemingly abnormal preoccupation of nearly all the natural philosophers during the classical period with meteoric phenomena generally. In the famous *De Rerum Natura* of Lucretius (96–55 BC) for example, one finds that the ordinary facts of astronomy or meteorology, the motions of Sun, Moon and planets, or the incidence of frost and wind, snow and ice, are not really of any great interest. The main focus of attention is the range of phenomena believed to be associated with divine agencies, namely comets and meteors, thunder and lightning, volcanoes and earthquakes, all regarded as supernatural events liable to induce a state of terror.[49] The Greek three-fold classification of comets into 'beards', 'cypress trees' and 'torches', several kinds of the latter clearly describing fireballs or bolide falls, is consistent with this view. This focus of attention, far from being anomalous, may be evidence of the enhanced fireball and meteoric flux during the classical period associated with the recent evolution of a large disintegrating comet or its orbit. It is true of course that by the end of the first millennium BC, many natural philosophers had finally settled on the wholly mistaken idea that comets and even the Milky Way derived from some kind of terrestrial whirlwind, but they have also left enough evidence to indicate that their imagined action of whirlwinds was probably no more than an explanation based on well-known meteor streams that extended into the sub-lunar zone.

For the most part then, the natural philosophers of the pre-Socratic era seem to provide us with descriptions of the astronomical environment in which the old Babylonian and Egyptian fears would not have been out of place. Indeed there is no convincing evidence that the model of the universe espoused at this time ever really disappeared. Certainly Lucretius and Seneca, much later on in the classical period, provide us with accounts of comets and meteors which are wholly consistent with these earlier views *and* in general agreement with our modern understanding of them. At the same time though, it is also clear that cosmological thinking underwent a profound change in ancient Greece and that this finally happened around the fourth century BC. A new tradition of planetary observation emerged, together with a new-found academic interest in geometry and circular motion.[50]

We might easily suppose the study of planetary movements over the sky, like the rise of the new rationalism which had no need of gods, was only likely to be pursued if the more awesome gods had been banished from the sky; and the tradition can in fact be traced back to the Pythagoreans in the sixth century BC. Indeed there is evidence (table 4) that the planets were first given names *after* this time though not, as it happens, the names of well-known gods. The latter were only tacked on later and it says something for the nature of the more ancient gods that the planets then acquired some rather artificial and quite inappropriate cometary characteristics. Even as late as the ninth century AD, the Baghdad astrologer Kitab al-Mughni described Jupiter as 'bearded', Mars as a 'lamp', Mercury as a 'spear' and Venus as a 'horseman' (plate 7). These terms may be compared with those used to describe comets by Pliny in his *Natural History*: mane, hair, beard, fleece, dart, sword, spear, cask, horn, torch and crown.

Table 4   Scientific and divine names of planets

Greek divine	Greek scientific	Latin divine	Babylonian scientific	Babylonian divine
Star of Cronos	Cronos	Saturn	Kaimanu	Ninib
Star of Zeus	Zeus	Jupiter	Mulu-babbar	Marduk
Star of Ares	Ares	Mars	Sal-bat-a-ni	Nergal
Star of Aphrodite	Aphrodite	Venus	Dili-pat	Ishtar
Star of Hermes	Hermes	Mercury	Gu-utu	Nabu

Simplified tabulation showing the history of planetary names in Greece, Rome and Babylonia. The Babylonian names are very old but cannot be precisely dated. The attachment of the names of gods to planets originated in Babylonia.
  Column 1: divine names used in Greece at the time of Plato (430 BC);
  Column 2: scientific names used in Greece in Late Antiquity (after 200 BC);
  Column 3: names of Roman gods attached to planets (after 100 BC);
  Column 4: Babylonian scientific name; association with planets possibly late;
  Column 5: Babylonian divine name.

The idea that the planets might be important only really came into its own however with the work of Plato and Eudoxus. The latter, in particular, elaborated the explanation of orbits in terms of steady circular motion whilst Plato emphasized the beauty and simplicity of the geometry that was implied. Plato's influence was considerable: he held that the world was ruled by logic and that its structure could be worked out by the application of pure thought. Even in the *Timaeus* for example the gods did not destroy Atlantis capriciously but through logical necessity. But Plato, as we have noted, still took note of the ancient world-picture.

The same can hardly be said of the radical Aristotelian cosmology that followed, however. Plato's pupil Aristotle (384–322 BC) took a more pragmatic view of the world and the ideal of science which he developed has carried immense weight even to the present day. As a practical zoologist his attitude was basically empirical, and he placed great emphasis on the more immediate evidence of the senses: 'here and now' rather than 'there and then'. The effect of this was to render received knowledge less credible, the more so if it was not in accordance with very immediate impressions. This then merely served to hasten the degeneration of cometary gods into distant folk memories, and in the cosmology which Aristotle developed, comets did indeed completely lose their status as celestial objects along with their portentous overtones. They came

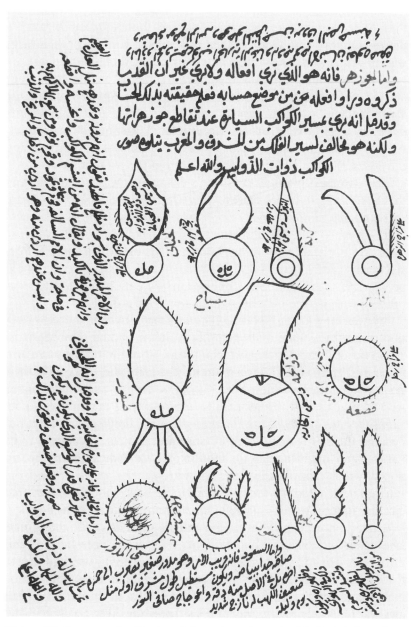

Plate 7. Ninth-century classification of planets from the work of the Baghdad astrologer Kitab al-Mughni, where each is given clear cometary characteristics. The descriptions used are: Top row, left to right: bearded (Jupiter); lamp (Mars); spear (Mercury); horseman (Venus). Centre row: giant (Sun); girl (Moon); kettle (Saturn). Bottom row: circle, lamp, skewer, jar, javelin (Princeton University Library).

then to be classified as somewhat ephemeral meteorological phenomena, no more significant than a passing shower. Opinion at large
was never convinced of this new-fangled mediocrity of cometary
gods and still looked on them with awe, but Aristotelian cosmology
and its sophisticated fascination with planets rapidly gained the
ascendancy in academic circles and, to the extent that state activity
depends a good deal on the accumulated wisdom of academics,
comets ceased to have any particular importance. To all intents and
purposes, this shift of view has persisted to the present day.

The strongest plank in Aristotle's cosmology was his assumption
that the Earth could not be revolving around the Sun. He was aware
that any such motion would appear to displace the stars in the course
of the year – the phenomenon of parallax. Since the immense
distances of the stars were not known in his day, this rationale for a
stationary Earth seems valid, but to keep it in perspective it should
be kept in mind that the Copernican revolution much later was
pressed without the benefit of this knowledge either!

Another new departure in Aristotle's teaching was the introduction of a fifth element called aether which occupied the outermost
space beyond the Moon and which was more divine than fire, air,
water and earth. From it were formed the stars and planets, in their
respective spheres, which were said to be eternal and intelligent as
well as divine. It was also a natural property of matter to seek its
rightful place in the world, for example earth down and fire up, but
because the divine material of the stars and planets was already in its
natural state of perfection, the motion of these bodies was neither up
nor down. By implication the motion was in circles around the
Earth, and it was another logical element of Aristotle's theory that
circular motion was more perfect than motion in a straight line.

Aristotle however did not regard the planets as we do today, as a
free-wheeling system; he thought they were interlocked and
ultimately driven by a Prime Mover whose sphere lay beyond that of
the fixed stars (plate 8); but the scheme, like similar ones which came
before and after, had many defects of which he must have been
aware. Thus planetary movements over the sky are somewhat
complex, being generally eastwards against the stellar background,
but sometimes looping in a retrograde direction; and their tracks lie
almost, but not quite, in the plane of the ecliptic defined by the solar
motion. To explain this complexity of movement Aristotle needed to
introduce 55 spheres for 9 objects (5 planets, the Sun, Moon, fixed

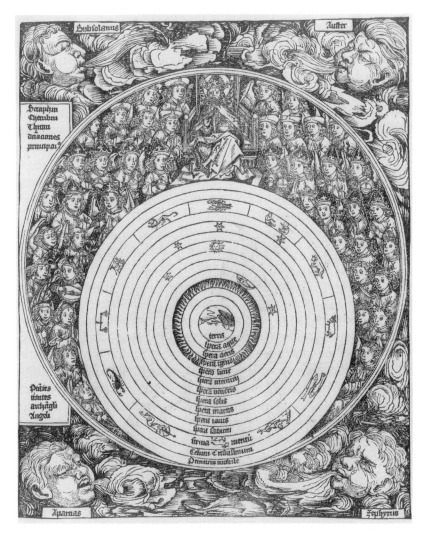

Plate 8. The Aristotelian planetary system as represented by the Nürnberg Chronicle of 1493 (reproduced by permission of Sotheby's, London). The Earth (terra) resides at the centre of a series of concentric spheres representing the other three basic elements (water, air, fire), all of which, according to Aristotle as he was interpreted by medieval philosophers, are associated with their particular ideal levels. Beyond that are the crystalline spheres to which are attached the celestial bodies: first the Moon, then Mercury and Venus, the Sun and the outer three planets, Mars, Jupiter and Saturn. Next comes the sphere of the zodiac and then the sphere of the fixed stars in countless numbers. The penultimate sphere is Aristotle's 'Prime Mover', whilst the final sphere is the domain of God and the Heavenly Host. This was to remain the popular image of the cosmos long after the investigations of Kepler, Copernicus and Galileo had posited a quite different order.

stars and the Prime Mover); an otherwise simple geometrical model, even from its inception, was an elaborate and highly confusing mechanism. Nevertheless, the disposition of matter in the universe, conceived as being perfect above the Moon and imperfect below, was obviously regarded by contemporary natural philosophers, as well as those who followed, as so elegant and reasonable that it transcended any detailed theoretical shortcomings. To be consistent with the Aristotelian scheme of things, therefore, there was further good reason that the erratic comets and the unpredictable fireballs should no longer be part of the perfect heavens. Indeed, it became necessary to dissociate comets from the gods altogether and relegate them to almost the lowliest plane of all.

Although Aristotle then had for reasons of pragmatism and theoretical elegance downgraded comets, and although there was, as we shall see, a political dimension to the 'planetization' of gods, these reasons only found favour at first in the Alexandrian school and were overshadowed by the simple fact that the cometary gods had disappeared. For as the sky gods faded, the myths repeated from one generation to the next would become less and less comprehensible. On the one hand the tales were plainly to do with celestial beings; on the other hand there was a problem of identity. As time went on, the only moving bodies left in the zodiac, apart from Sun and Moon, were the planets. The planets were too few in number and too simple in behaviour to carry the rich complexity of myths; but they had become the only candidates available. Inevitably then, the transference of the celestial adventures of the gods from comets to planets, for which the Greeks were primarily responsible between 600 and 200 BC, extended throughout the civilized world.

The transition also marks an interesting change in the perceived nature of the gods. Henceforth they appear more abstract, remote and impersonal, so much so that some commentators remark that they are no longer the typical objects of a religious man's worship. This is not to say that Aristotle was basically a secular thinker however for there are references to living gods to be found scattered throughout his writings. The presence of these new characteristics is more an indication that Aristotle was concerned to remove the elements of fear and superstition from man's relationship with gods. He dismissed the old anthropomorphic deities of the Olympian pantheon as mere myths and skilfully covered his tracks by also insisting that 'our ancestors' did not purvey unmixed superstition, having dimly perceived that a certain primary substance was divine and that this primary substance was to be found in the heavens. Such skill enabled Aristotle to maintain a world-view that did not fly in the face of the ancient divinities but nevertheless made it possible to ascribe, to the new-found planets, supernatural powers that eventually went far beyond their inherent capabilities.

There was perhaps a degree of artifice in this line of argument, but one should keep in mind the prominence given to open confrontation and debate in the society of ancient Greece[49]. It has been said that:

the institutions of the city-state called for new qualities of leadership, put a premium on skill in speaking and produced a public who appreciated the exercise of that skill. Claims to particular wisdom and knowledge in other fields besides the political were similarly liable to scrutiny, and in the competition between the many and varied new claimants to such knowledge, those who deployed evidence and argument were at an advantage compared with those who did not. . . . Examination of the Greek evidence suggests that this very paradigm of the competitive debate may have provided the essential framework for the growth of natural science.

But if this is so, then it was words rather than observation, logic rather than experiment, which lay at the heart of Greek science; and we can be reasonably certain that it was Aristotle's intention, as well as his achievement, to mesh revolutionary astronomical theory into a pattern involving new gods that most of his contemporaries would be unable to dent by argument. Right from the start therefore, the new planetary gods were provided with a secure rationale based on perceived fact. But it was barely possible to have the one without the other: Aristotle's new cosmology demanded wholesale commitment to the new, tamed gods in a state of heavenly perfection.

There was a political dimension also to the revolution. What, now, was the role of the gods in human affairs? The dissension over this issue went back to the very foundation of the Athenian assembly, and it both fuelled and was fuelled by the lack of cohesion amongst the city-states which frequently threatened to return Greece to the conditions of the Dark Age. Inevitably, there were those who had an interest in the orderly management of affairs and who saw definite advantages in promoting the new anodyne version of the gods. As Karl Marx was later to recognize, there was a sense in which the more sophisticated aristocrats of ancient Greece may have perceived the potential benefits of treating religion as the opium of the people. This attitude may already have been implicit in the paternalism of the old Egyptian state but if something like the cosmic pharaoh could not be reintroduced, sophistry would certainly permit the trappings of religion and public debate to continue whilst a privileged few, committed only to neutral gods, manipulated the machinery of government and Academe behind the scenes.

The aristocratic Critias allows us a glimpse of these ideas in the making when he suggests:

> that a man of shrewd and subtle mind invented for men the fear of the gods so that there might be something to frighten the wicked even if they acted, spoke or thought in secret. To this end he introduced the concept of divinity. There is, he said, a spirit enjoying endless life, hearing and seeing with his mind, exceedingly wise and all-observing, bearer of a divine nature. He will hear everything spoken among men and can see everything that is done. If you are silently plotting evil, it will not be hidden from the gods, so clever are they. With this story, he presented the most seductive of teachings, concealing the truth with lying words. For a dwelling he gave them the place whose mention would most powerfully strike the hearts of men, whence, as he knew, fears came to mortals and help for their wretched lives; that is, the vault above, where he perceived the lightnings and the dread roars of thunders, and the starry face and form of heaven – fair wrought by the cunning craftmanship of Cronos; whence too the burning meteor makes its way, and the liquid rain descends on earth. . . . So, I think, first of all, did someone persuade men to believe there exists a race of gods.

It is not to be supposed that these developments emerged in their final form overnight. Plato and Aristotle were regarded with no great awe in their own lifetimes even though it was their explorations into the realm of new ideas that ultimately broke the ancient traditions. Nor must it be supposed that there was some kind of knowing conspiracy between the aristocracy and Academe. Of course, it often happens that some particular way of looking at the natural world finds itself in harmony with some particular way of ordering society. For example, in a long-established monarchical state, it is easier to see the natural world in terms of design and divine government; on the other hand, in the ambience of a non-authoritarian democracy, a natural world of order and authority might seem less plausible than one of materialism, without divine intervention or design.

The assignment of cause and effect in such a situation is bound to be an elusive matter. Nevertheless, with the benefit of hindsight, we can now see that it was almost certainly Aristotle who, breaking away from his mentor Plato, dissociating comets from the celestial realm, and invoking purely circular motion for the planets on Platonist grounds of elegance, appreciated that there could never be any question of the planets coming into contact with the Earth. In principle therefore, the Earth was safe and there was no ultimate sanction on degenerate people by gods in the form of a cataract of fire. Public opinion regarding the authority of sky gods might not be gainsaid but it could be diverted, and more or less inevitably, Aristotle's Prime Mover became identified with the principal Deity whilst the new planetary gods were regarded as exerting quite illusory 'planetary influences' across empty space, these influences controlling events here on Earth. There therefore arose the essence of a world view in which horoscopic astrology seemed perfectly natural.

In this way, out of a very basic need on the part of rulers to invent gods who corresponded as closely as possible to an orderly public perception, several generations of Greek academics moulded a brand new cosmology which was subsequently perfected by the Alexandrian school and came to serve as the basis for horoscopic astrology. It is a matter of history now that it took subsequent generations almost two thousand years to unravel the complexities introduced by the Aristotelian style of argument, and to recognize the fundamental importance, for the purpose of arriving at the truth, of dissociating the arguments of Academe from the requirements of crowd control in the face of a perceived danger. It is often not appreciated nowadays how this fundamental requirement had to be learned the hard way.

Through an extraordinary combination of intellectual devices and errors then, during a period of declining comet activity and diminishing zodiacal cloud, which was nevertheless enhanced above current levels, Pythagoras, Plato and Aristotle led their contemporaries and indeed their successors to a seemingly plausible and yet wholly erroneous view of the nature of the universe, based very largely on a contrived understanding of the behaviour of the planets.

Of the various states which engaged in the war of 371–362 BC, Athens, with the exception of Thebes, fared the best. However, the actions of the Athenian government evidently became increasingly arbitrary. Samos, for example, which had fallen into the hands of the

Persians, was conquered in 365 BC but instead of freeing her old ally, Athens established an island colony by exporting a large number of her poorer citizens. Those other towns which had come together in league with Athens in 378 BC looked on in disapproval. So too did the fiercely independent Macedonians who, though not very distant kinsmen of the Greeks, continued in the fourth century BC to live a kind of tribal life not very different from that of the Hellenes at the end of the Dark Age. The Macedonian nobility were familiar with the Greek language however and had adopted Greek names for themselves such as Archelaus, Pausanias, Lysinachus and Ptolemaus, as well as for their gods. Foremost in the Hellenization of Macedonia had been their king, Alexander I, who invited Socrates to Pella to instruct the youth. And it was his great grandson Philip II (359–336 BC) whose inspiration now began to turn the tide for the Hellenes.

Philip II was an unscrupulous operator who both admired and despised the Greeks and their culture. He obviously took a pleasure in outwitting his Athenian adversaries and nothing stood in his way when it came to advancing Macedonian interests. Corruption and the sword his favourite weapons, Philip rapidly dominated Greece and brought Athens within his grasp. He also brought the great philosopher Aristotle to his court in order to be the tutor of his 13-year-old son. And the omnivorous appetite for knowledge which inspired Aristotle and ranged over every subject from botany to metaphysics and from constitutional history to moral philosophy, seems to have influenced the young Alexander in no small measure. A clever, inquisitive restless Greek mind was thus cultivated in a rather special Macedonian pupil by a rather special teacher.

Upon the assassination of Philip, Greece knew little of the young man who came to the Macedonian throne. But it was soon clear that the power of Philip had fallen into the hands of a man even greater than Philip himself, one of the most extraordinary characters that Europe has ever known, and a man whose personality was to be stamped on the history of the world for at least a thousand years. There were other strains in his character besides that of the generous knight; he had a strong infusion of the unscrupulous energy of Philip: those who crossed his path or merely incurred his suspicion he swept away without pity or remorse; and as he grew older, he became as conscienceless as his father. But the quality which enabled Alexander to leave his mark on history was his military talent. He was a heaven-born general, and was besides brought up with every advantage that he could have desired.

The moment that Philip was dead all Greece gave a sigh of relief and prepared to forget the Macedonians and recommence their usual intrigues and wars. But before anything more serious was done, Alexander swooped down among them with thirty thousand men at his back, compelling the unprepared Greek states to elect him as supreme commander of the Hellene confederacy. Within six months he was back in Macedonia preparing to deal with incursions along the northern frontier and assaults on the force which his father had sent into Asia across the Hellespont. These latter were a source of particular annoyance and nothing could be more inspiring to the enthusiastic mind of Alexander than a crusade of Hellenism against the Persian barbarians to the east. In the event, his invasion of Asia Minor was followed by an early strategic withdrawal by Darius and this simply inspired even greater aspirations. Anatolia, Levant, Palestine had all fallen to the conquering Greeks by 332 BC whilst Egypt capitulated without a blow: its inhabitants regarded the Macedonians as deliverers from the Persian yoke against which they had long striven. Alexander made a triumphant entry to Memphis and then sailed down the Nile to its western mouth where, struck with the capacities of the spot, he drew out a plan for the foundation of a great maritime city and christened it by his own name. Thus came into being the seaport of Alexandria, one of the most enduring and famous of all the monuments which Alexander reared for himself.

Whilst staying at Alexandria, the new king resolved to visit the famous oracle of Zeus Ammon in the Libyan desert. With a hand-picked regiment of troops he marched for five days across the sands, and came in safety to the palm groves of the fertile oasis which sheltered the temple of god. The oracle wisely hailed him as the son of Zeus, and bade him go forth and conquer all the world, for none would be able to withstand him till the day when he should be taken up to the gods. His company were then bidden to salute him as more than a mortal, and to offer him sacrifice. This hyperbolical flattery seems to have turned the head of Alexander: it was noted that he took the oracle in all seriousness, and was in future much pleased when anyone saluted him as the son of Ammon.

In the spring of 331 BC, Alexander retraced his steps through Egypt, Palestine and Syria, crossing the Euphrates and the Tigris and advancing into Persia. By now, no one doubted that he was the 'great king', Darius being no more than a luckless pretender, and the Persian realm soon fell. Meanwhile, Alexander's pride and vanity knew no limit; he took to assuming divine honours as his right, dressed himself, to the surprise of his comrades, in the purple robe and tiara of an eastern king, and surrounded his person with oriental courtiers. But by the time of his early death, Alexander had completely transformed the outlook of the Greeks. The Macedonian conquest of the East revolutionized the relations of the little Hellenic states both with each other and with the outer world. The old system of local autonomy and constant wars to maintain the balance of power had now become impossible. What could be more inspiring than to see the old Hellenic genius for colonizing was not extinct; to behold the conquerors laying hands on every province from the Aegean to the Indus, and covering them with Greek cities as great and as vigorous as any that had existed in the Hellenic fatherland? In due course, under the firm management of one of Alexander's generals, Ptolemy Soter, who ruled from 305 to 283 BC, the city of Alexandria became the seat of the new Hellenism. The famous Museum and Library of Alexandria were set up under Ptolemy, and the scientific enterprise shifted from mainland Greece towards Alexandria.

However, as the customs of local autonomy familiar to the Greeks gave way to those of a more centralized regime, the head of state was able to take on many of the trappings of a divine, cosmic ruler, a type familiar to the Egyptians. There was a difference between the new cosmic ruler and the older archetype, however, for this was no longer a head of state conscious of gods exercising an ultimate sanction in the form of cleansing cataracts of fire. The sanctions now exercised were mere planetary influences, and invented ones at that. If these were to be effective, the interests of state religion, public order and Academe clearly had to be intertwined. This may have assisted the purposes of government but questions of natural truth were now vulnerable since they risked becoming matters of appearance for the sake of public order.

In fact, it is now known that the scientific purpose of the Aristotelian system in the hands of the Alexandrian school which followed was to 'save appearances',[51] and that it also had theological connotations which, if Strato's stage-managed miracles are not misunderstood, were attributable to a wholly invented divine control. Strato invented the science of pneumatics, his knowledge of which he exploited to counterfeit, in a variety of mysterious ways, the turning of water into wine and the mysterious movements of idols across the altar, these and other effects being designed to deceive the worshipper. It is no accident therefore that in trying to account for the motions of the planets over the sky, geometrical models were constructed with little or no concern for whether they might be true!

The continuing failure to find a theory of the planetary system had far-reaching effects of which the most serious was a strengthening of the theological aspects of astronomy. Where Aristotle himself may well have wished to treat the universe as a purely mechanical system with only a Prime Mover as the solitary theological element, under the influence of the Alexandrian school, every aspect of the heavens tended to become an object of worship. Now, however, the heavenly gods were not comets but planets. Moreover the critical phases of human existence came to be ascribed to a remote action on the part of these 'new' planetary gods, which was elaborated in terms of the quite imaginary theoretical structure now known as horoscopic astrology. Horoscopic and omen astrology were indeed quite different. Whereas the latter began as an honest attempt to anticipate the effects of real encounters between meteor streams and our planet, the former was an attempt in the name of science to confer realism on an imagined planetary process. Horoscopic astrology eventually found its highest expression in the works of the Alexandrian scholar Claudius Ptolemy (AD 100–165), known as the Almagest and the Tetrabiblos, and it is a matter of record now that these works continued to appeal to the enlightened minds of western civilization for most of the subsequent fifteen hundred years, first within the Islamic empire and then in Renaissance Europe.

Scientific commentators given to seeing the growth of science as an exclusively mathematical tradition have succeeded in having the Tetrabiblos, which is a systematic exposition of astrology, regarded nowadays as an aberration outside the main stream of science. But, of course, the physical picture was fundamental; and it was not until the basically irrational character of wholly invented astrological forces 'acting at a distance' had been uprooted by the Cartesian philosophers sixteen hundred years later, in the wake of the exposure of Aristotle's various misconceptions by Copernicus, Bacon, Tycho, Kepler and Galileo, that any major new advance in science and mathematics could be made. It is well understood now that the long dominance of horoscopic astrology, and the slow progress of science over this time, has its roots in the failure of the Greek experiment with rationalism; what has not been appreciated, until recently, however, is that in creating the planetary gods, the Greeks also diverted attention from the catastrophic actions of the real celestial gods as expressed in the oldest tales.

By the seventeenth century, despite misgivings on the part of the Church, it was accepted amongst astronomers that there was a definite need to simplify the system of concentric crystal spheres carrying the planets and that it was necessary to put the Sun in the middle. In order to explain the complexities of planetary motion, Descartes replaced the whirling glass spheres with a Sun-centred system of transparent fluid vortices in the heavenly ocean, an idea which Newton subsequently endorsed and sought to express in precise mathematical form: a theory for the planetary system did then at last arrive.[52] From this juncture the writing was on the wall for horoscopic astrology; further, with comets again in the celestial domain and perceived as an impact hazard, there was once again a basis for omen astrology and its catastrophist perception of the oldest myths.

Indeed, as we shall see, the arrival of two bright comets in 1680 and 1682 almost shattered the carefully nurtured Aristotelian dream, and there followed a period of considerable intellectual ferment. Newton however saw a way out of the difficulty, and the brief opportunity for a new understanding of ancient history passed. In its essentials, Newton's new-look Aristotelian scheme, in which the Earth moves untroubled by cosmic forces, has survived to the present day. However, before we reach this phase, we have to understand how it is that the Christian Church took on the Aristotelian banner.

7

Doomsday

THE spasmodic disintegration of a particularly impressive comet, then, could well have been responsible for a series of disasters in the Mesopotamian and Egyptian civilizations during the third millennium BC, and then again for a subsequent series of disasters inflicted upon the Minoan and Mycenean civilizations during the second millennium BC. Catastrophes of this sort, delivered by visible celestial gods, are completely outside modern experience, but it is clear that they could have been a major reason for the preoccupation with, and dread of, the sky manifested by the earliest civilizations.

Conversely, by the middle of the first millennium BC (around the time of Socrates), when these encounters had become less frequent, the capacity of the visible cometary debris to do any harm must have been called into question. It is plausible to suppose that in these more quiescent circumstances a new outlook on the world would come to the fore, one familiar to the modern era, in which moral sanctions were now provided by civilizations themselves rather than by way of thunderbolts hurled by the gods. Amongst those so committed, the tendency must have been to assume that the world was safe from celestial visitation.

At the same time old ideas die hard and it is also plausible to suppose that the fundamental change of world-view enforced by the appearance of the sky was not instantaneous and that it became a source of conflict and debate. Indeed if the timescale of the debate were comparable with that over which celestial activity might revive, say with cometary encounters or fireball activity reaching some kind of peak during the centuries immediately before Christ (see figure 4), such conflict and debate would be intensified, simply delaying for a while any widespread conclusion that the world is safe.

This account of the course of cosmic events places the pre- and post-Socratic periods at Athens in a new light. Thus it is possible to detect some degree of exasperation in Athens with the failure to

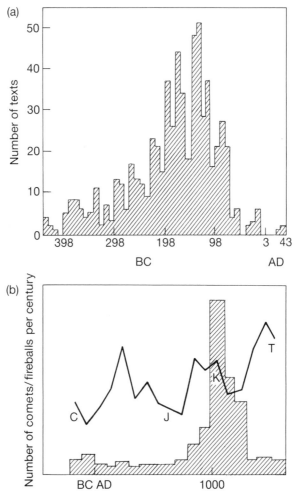

Figure 4. Diagrams illustrating the apparent enhancement of fireball activity (a) during the late Babylonian period (based on recovered ephemerides and monthly records on cuneiform tablets); and (b) during the Chinese T'ang dynasty (based on recorded fireball and comet observations, where the commonest fireball radiants during the eleventh-century peak arise from the Taurid meteor stream). In interpreting these apparent surges account must be taken of variation in the attention given to the sky by astronomers at different epochs (see also *The Origin of Comets* by M. E. Bailey and the authors). Also relevant is the location of astronomers, the increasing trend in the observed number of comets per century apparently reflecting the spread of professional astronomical practice from the Chinese (C) to the Japanese (J) and Koreans (K) before the introduction of the telescope (T).

relinquish the old, mythical gods, so much so that Macedonian aristocrats were caused to embrace a new, more impotent supernatural power and impose their will upon a vast new (Hellenic) empire. Indeed the particular new view of the world which the Macedonians espoused, in which the gods were a kind of image to be exploited to the advantage of the state, almost certainly flourished only because of their ability to defy their homeland and carve out their own 'brave new world'. Further, it was only a matter of time before the rapidly expanding Roman Empire engulfed Greece and Egypt and itself absorbed the new 'world-view'.

It may be no coincidence that the periods immediately after Socrates and Christ – after 450 and 0 BC respectively – saw the eventual flowering of academies in Athens and Alexandria respectively, which sought to provide an understanding of the natural world within the constraints of a selected 'world-view'. In each case, there was a tendency to enforce the chosen world-view by strict adherence to particular party lines. The centre of learning at Alexandria, for example, right from the start, had a certain transatlantic opulence, reflecting its status in this brave new world, and a deference to the agreed cosmological view which was bound up with the stability of a state controlled by a ruling oligarchy. To hold alternative views was therefore to risk not just scientific error, but heresy. The state increasingly assumed a new inquisitorial role.

The pressure to conform with the new Aristotelian world-view was therefore very strong, and we may safely assume that most unwanted history and ancient knowledge was relegated to the mythological dustbin. No longer was there any significance in Atlantis, no longer did bodies exist revolving in heaven that brought destruction to the Earth at long intervals by a great conflagration. Eliminated also was any vestige of the old dualism in the sky between the forces of good and evil. All that remained eventually was an old benevolent Zeus, underscored by Aristotle's principle of order and harmony, the Prime Mover, contending with a malevolent 'necessity' invented by Plato and located here on Earth. According to theory therefore, all movement and good had to be attributed to the Prime Mover whilst sloth and evil were to be counted as fundamental aspects of matter and human life, as necessary elements of the lowest level in the cosmos. This combination of evil with human nature emerges now as a fundamental notion, creating a facet of theology that seemed to be in excellent accord with human experience. Indeed, the specially invented new religion now emanating from Alexandria, the so-called cult of Serapis,[53] was able to present itself in a new guise as an agent of reform, an active missionary force on the side of good. In the hands of an Alexandrian trained priesthood, armed with its pantheon of planetary gods and stage-managed miracles, the temples of Serapis spread widely throughout the Near East and looked set to embrace the world!

It seems however that the cult of Serapis displayed rather too obviously its aristocratic connections; it was perhaps too concerned with imposing good habits and not enough with offering reward or hope in a frightening world. In this respect the concern was still with the seasonal flux of meteors and fireballs, and the anticipation of catastrophe and world end. Prophets of doom, as of old, were still prevalent. On the other hand, there were the commissioners of organized religion who practised horoscopic astrology and invoked the new Aristotelian doctrines based upon the supposed influence of the planets. Their concern was more with calming public fear and expressing the essential good will of God. Based, it seems, on formulae of apparently proven success, both the trappings of organized religion and the fundamentals of theology throughout the Near East seemed to develop a certain degree of uniformity: monotheism, astral influences, miracle-working and so on.

Out of this melting pot, there emerged some kind of pattern, a division of belief in the community between horoscopes and omens polarized more or less along class lines. The upper and middle classes seem to have favoured the more cerebral Hellenistic cults whilst the lower orders remained with the more traditional beliefs. This division may well explain why the struggle for supremacy persisted for so long, but we must not overlook the influence of the sky:[54] it is well known to us through Chinese records, also supplemented by those of the Babylonians (figure 4(a)) and the writings of classical authors, that there was a significant enhancement of meteoric activity towards the end of the first millennium BC, starting perhaps around 250 BC and continuing for some decades into the following era as well. There were, it seems, markedly increased expectations of an impending world-end at this time which continued unabated for much longer than a century.

It was amongst members of the Christian cult that the expectation of world-end emerged eventually with the greatest force. To follow this development, we have to examine the evolution of ideas in one of the countries closest to the centre of the Hellenic empire, namely Judea.[55] Originally Judaic cosmology exactly mirrored that of Greece and Egypt and was probably of Mesopotamian origin: here too there was a flat earth with 'waters' above and below (figure 5); here too there had been a great struggle in the sky between deified figures, Behemoth and Leviathan; and here too, the dualism had evolved until in the mid-second millennium BC, it was a well-intentioned Yahweh and a not-so-well-intentioned Satan who were in contention. However, if there was a difference that eventually set Judaism apart from other religions in the Near East, it was in the fierceness of the attachment between Yahweh and his devotees. This variation on an otherwise common theme arose amongst the Jews at the time of their escape from Egyptian bondage when certain cosmic forces were said to have wreaked destruction upon their captors.[56] Henceforth the Jews regarded Yahweh as their particular protector and themselves as His chosen people.

The special relationship was deepened by the intense isolationist policy of subsequent leaders. Nevertheless by the closing centuries

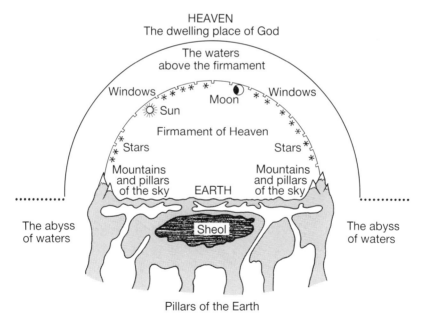

Figure 5. Representations of the Judaic cosmos. Note the great arch representing upper and lower heavens, the latter corresponding to the path of the Sun and Moon. The windows in this path appear to be associated with store-houses for the winds, upper waters and hail (line drawing reproduced by kind permission of Weidenfeld & Nicolson).

of the previous era, doctrine had divided, as it had at Athens and Rome, along more or less devil-may-care (epicurean) and severe doom-laden (stoic) lines, represented locally by the Pharisees and Sadducees. Under the liberating influence of the Hellenic and Roman domination however, there also emerged many prophet-led sects of which the Mandean, Gnostic, Essene and Hermetic seem to have been the most significant. Later, all reference to the teachings of these sects was to be systematically suppressed, so it is difficult now to get a proper appreciation of their importance; authorities like Josephus and Pliny for example, in the first century AD, gave them only passing mention. Each sect however was characterized by esoteric doctrines including the customary angelology, eschatology and (omen) astrology, as well as a sacred literature.

The Essenes in particular appear to have drawn their doctrines from a Mesopotamian source, bringing with them a Zoroastrian dualism in which man was affected in equal measures by the sons of Light and Darkness. Zoroastrianism was an important near-Eastern religious movement in the middle of the first millennium BC, ultimately of Iranian origin. Led by 'magi', it subscribed to Ahura Mazda as the sole, benevolent creator of the Universe. It also postulated primeval twin spirits, Spenta Mainyu (Good Mind) and Angra Mainyu (Evil Spirit) with their respective angelic and demonic followers in eternal conflict since the dawn of creation, and so had many of the now familiar characteristics of older religions. There were differences of emphasis, of course, and different names for the gods, but there was plainly something of a common background between the Judaic and Zoroastrian faiths.

However, the specific role which the Essenes took upon themselves was the study of ancient prophecies relating to the immediate future, which historical events had convinced them were the long-predicted Last Times, not only the culmination of world history but the antecedent of an incipient Golden Age. Thus convinced, they produced a plan of action, or book of oracles, heavily laden with astronomical detail, describing:

exactly how and in what order the sect must address itself to deliberately bringing about the fulfilment of all the outstanding biblical prophecies. In this way, all the conditions would be fulfilled for the inauguration of the long-awaited kingdom. Everything therefore hinged upon the prophetic interpretation of the individual teacher and his successors. With its aid there now seemed to be every prospect that beleagured mankind could at last take the promised kingdom of heaven by storm.[57]

As is well known, the belief that Jesus of Nazareth brought this astronomically oriented programme to fruition was eventually advanced with considerable success by the Christian sect of the Essenes. Their argument that his birth and resurrection were attended by heavenly portents is evidently not without significance, particularly when one has regard to their special claim that Jesus was possessed of divine qualities during his sojourn on earth. However the posthumous failure of the Kingdom of Heaven to arrive was open to misinterpretation and the later achievements of the Christian sect owed much of their success to a subsequent reinterpretation of the original programme by Paul of Tarsus. Taking his inspiration apparently from a sign, which may be interpreted according to taste as the occasion of a violent fireball or an epileptic fit, Paul modified the Christian message along Aristotelian or Hellenic lines by eliminating the anticipated conflagration and world-end.

Whatever significance we might like to place upon the Pauline revelation, it is clear that one particular Aristotelian notion, that meteors and comets were of terrestrial provenance, did not at this time have general support. There was still a tendency, with which Christian cosmology now came to be associated, to return these objects to the celestial sphere. Not as a potential source of destruction however, as the omen astrologers assumed, but as purposeful signals from heaven. Paul thereby dissociated the earlier messianic expectations from a catastrophic world-end and linked them instead with a spell of suffering by the Son of God, identified as Jesus, during a predicted return to Earth. As we have already noted, it was an old Babylonian idea that kingly suffering was necessary for the sake of humanity as a whole. Admittedly there were difficulties in accounting for many of the apocalyptic statements in the sacred literature but there was usually sufficient ambiguity in a cosmic messiah 'who takest the clouds for thy chariot, riding on the wings of the wind; who makest the winds thy messengers and flames of fire thy servants' not to undermine the revisionist thinking.

In this way Paul's skilful blend of Judaism and Hellenism successfully overcame the principal failing of Aristotelian theology, namely its inability to explain the apparently astronomical signs which were a source of despair to so many. By relating these signs to the temporary presence of God on Earth, however indispensable such an idea might appear on other grounds, the basis for confidence in the future and hope, which already characterized Hellenism to some degree, was strengthened still further. As a result, Christian theology came to be seen by managers of State and Academe alike as providing the soundest possible theoretical basis for natural, civil and moral law. Through the eventual good offices of the Holy Roman Empire, associated with divinity through the papal succession, the Christian sect multiplied and during the first millennium AD became the dominant force in Near Eastern and European religion.

And so, during a period of reduced meteoric activity in the first few centuries of the present era, Christian theology began to be licked into shape. If we examine the acrimonious and frequently violent conflict that ensued, as different factions struggled for dominance, it was the question of the divinity of Jesus that naturally emerged as the most pressing issue. Any deviations from this basic principle, which might have reduced Jesus to the level of a mere prophet (with special skills in astrology!) were ruthlessly expunged. The dominance of the Christian Church indeed came to rest more and more assertively on the presumed divinity of Jesus and the continuing association with this divinity through the papal succession. How this succession was to work was hardly self-evident; but it was a principle that any government involving an Athenian-styled 'ecclesia' needed to insist on if it was to claim power through the sort of divine or cosmic association that the Egyptian or Hellenic monarchies had previously passed on through a dynastic succession. The process now takes place behind closed doors and seems to be crucially attended by a puff of smoke!

It is interesting that even this issue may not have been settled automatically since there is some controversial evidence that the Holy Roman Empire itself was forged only by absorbing and suppressing various claimants to a Christian dynastic succession.[58] Be this as it may, the survival of Christianity could never be regarded as certain and much of its capacity to survive through the second millennium AD has depended on the outcome of several theoretical conflicts during the first millennium AD. For as Christianity spread into northern Europe, it came increasingly in contact with various forms of paganism each, as usual, carrying its own angelology, eschatology and omen astrology developed to varying levels of sophistication. Maybe these cults would have capitulated more rapidly to Christianity, as others had done before, had it not been that the sky degenerated again during the first millennium AD.

The evidence then is of a Christian Church during the early centuries AD that increasingly adopted a doctrine of the kind promulgated by Paul as it moved closer to the sources of political power. However, the continuing importance of the apocalyptic tradition in early Christianity, millenarianism as it has come to be called, should not be underestimated. Thus, apocalyptic ideas did not simply die out; they persisted, and indeed by the end of the fourth century AD they reached new levels of intensity throughout the whole underworld of popular religion. It was largely due to these undercurrents that the idea of the Saints of the Most High was revitalized to become a new potent force in Christian circles just as the Patriarchs of old had among the Jews. Also the idea of divine election re-emerged and entered the Christian apocalyptic as well. Indeed, the body of literature inaugurated by the Book of Revelation encouraged many Christians to see themselves as the Chosen People of the Lord, that is, chosen both to prepare the way for and to inherit the millennium.[59]

A contemporary expression of these ideas leaves one in little doubt as to what was envisaged: it was said that an Antichrist,

raging with implacable anger, will lead an army and besiege the mountain where the righteous have taken refuge. And when they see themselves besieged, they will call loudly to God for help, and God shall hear them, and shall send them a Liberator. Then the heavens shall be opened in a tempest, and Christ shall descend with great power; and a fiery brightness shall go before him, and a countless host of angels; and all that multitude of the godless shall be annihilated, and torrents of blood shall flow. . . . When peace has been brought about and every evil suppressed, that righteous and victorious King will carry out a great judgement on the earth of the living and the dead, and will hand over all heathen peoples to servitude under the righteous who are alive, and will raise the (righteous) dead to eternal life, and will himself reign with them on earth, and will found the Holy City, and this kingdom of the righteous will last for a thousand years. Throughout that time, the stars shall be brighter, and the brightness of the sun shall be increased, and the moon shall not wane. Then the rain of blessing shall descend from God morning and evening, and the earth shall bear all fruits without man's labour. Honey in abundance shall drip from the rocks, fountains of milk and wine shall burst forth. The beasts of the forest shall put away their wildness. . . . [and so on].

There are fanciful elements, of course, in this millennial vision, but one is bound to wonder whether some underlying cosmic spectacle was causing it. As Cohn points out, the forthcoming millennium was so widely attractive 'that no official condemnation could prevent it occurring again and again to the minds of the under-privileged and oppressed', and the institutionalized Church needed to display 'the utmost skill in controlling and canalising the emotional energies of the faithful, particularly in directing hopes and fears away from this life towards the next'.

Augustine in particular, during the fifth century, was to propound a view of the millennium which later became official doctrine. In his *City of God*, he argued that the Book of Revelation could be under-stood as spiritual allegory whilst the millennium had already arrived with the birth of Christianity and was indeed realized in the exist-ence of the Church.[60] He also presented the Church as a defender of ancient tradition and as the sole dispenser of divine essence (grace), thereby guaranteeing a formal role for the Church hierarchy and its establishment, both maintaining an essential cosmic connection and justifying the extension of its mission by increasing the 'card-carrying membership'. Nevertheless the fourth and fifth centuries remained a period of deep schism within the Church, revived paganism outside and intensified monastic withdrawal. The evi-dence points all too clearly to human conduct as a matter for general concern, driven it would appear, by a growing fear that the heavens would indeed open up in a tempest. Inevitably the rule of the estab-lishment was either blamed or condoned, abandoned or enforced, but the social fabric was torn apart all the same: an age of massive migration ensued and a more or less total lapse of central control. Barbarian incursions grew relentlessly, slowly but surely sapping the strength of the Roman Empire until in 410 AD, Rome itself was to fall ignominiously to the Visigoths under Alaric. By this time however, Europe was on the brink of another Dark Age.

One of the most extraordinary documents of the Dark Age, written some 130 years after the sack of Rome and the collapse of Roman rule in Britain, is a tract due to the British priest Gildas. Entitled 'De excidio et conquestu Britanniae', it describes in broad historical outline the course of events leading to what is called 'the ruin of

Britain'. The tract is not primarily a historical document however since Gildas's main purpose was clearly to admonish his contemporaries for standards of behaviour that left much to be desired, using recent past events to illustrate the likely course of heavenly retribution to come. Indeed, as we shall see, the contemporary period was one of strongly recurring meteor activity while Gildas, greatly respected in his own time as a sage and by historians such as Bede in centuries that followed, is now usually regarded as the *only* contemporary writer who attempted, however inadequately, a general survey of British history from the latter days of Roman rule until the middle of the sixth century AD.[61]

Remarkably, Gildas shows little interest in any part played by Anglo-Saxons in the early stages of the collapse of Roman and sub-Roman authority, and indeed, when Saxons appear, it is not as devastating raiders and invaders of Lowland Britain but as federate allies settled there by the leader Vortigern to help defend the land against incursions from the Picts and Scots. Furthermore, following the changes in official doctrine introduced by Augustine, the Roman Church was obliged to take very seriously certain heretical tendencies in Britain, and we know from accounts of the visit to Britain in the period before 440 AD by Germanus, the bishop of Auxerre, to discuss these problems, that sub-Roman Britain was still a force to be reckoned with and that Vortigern was in charge of a fairly well-ordered regime. Soon afterwards, however, it is clear that this regime suffered a quite exceptional reverse, the disaster being felt the length and breadth of the land. The outburst, which was 'the ruin of Britain', was marked apparently by widespread slaughter and destruction. In this episode, whose duration is not specified, Gildas deplores in particular the physical overthrow and ruin of towns, and the massacre of Christian clergy, both decisive in bringing about a virtually total liquidation of the most characteristic features of late Roman urban civilization.

The problem for historians[61] is that Gildas writes of all this in melodramatic phraseology 'employed deliberately', it is said, 'to create the maximum impression'; yet it is also clearly recognized that Gildas must here be treated as a witness of truth, for 'it would have been fatal to his argument and the aims of his tract to have described in such terms a situation which his readers knew to have been quite different'. Gildas reports on the events leading up to the catastrophe of the 'feathered flight of not unfamiliar rumour' (the rumble of incoming meteors, one wonders, reduced to subsonic speed), and of the catastrophe itself that it was due to:

the fire of righteous vengeance, kindled by the sins of the past, [which] blazed from sea to sea, its fuel prepared by the arms of the impious from the east. Once lit it did not die down. When it had wasted town and country in that area, it burnt up almost the whole surface of the island, until its red and savage tongue licked the western ocean. . . . All the greater towns fell to the enemy's battering rams; all their inhabitants, bishops, priests and people, were mowed down together, while swords flashed and flames crackled. Horrible it was to see the foundation stones of towers and high walls thrown down bottom upwards in the squares, mixing with holy altars and fragments of human bodies. . . . there was no burial save in the ruins of the houses, or in the bellies of the beasts and the birds.

The rhetoric, for such it is commonly deemed to be, leaves little to the imagination. On the other hand, the destruction in the towns, of which Gildas speaks, was still evident a century later; there was a desperate call to Aetius, the Roman leader in Gaul, for help; there does appear to have been an official delegation to Britain soon after, led once again by Germanus; and, subsequently, an exodus across the Channel on such a scale as to justify the future name of Brittany. The population of Britain did indeed plunge and the archaeological record is one of extreme deprivation and bare subsistence living for a generation or more thereafter. The scale of the disaster is not really in doubt, therefore; and as Myres[61] has pointed out, the evidence in the future spoken language and in place-names over a substantial chunk of Britain, extending from the Midlands to East Anglia, demonstrates that the previous Roman and sub-Roman civilization was completely wiped out. The evidence in the soil, moreover, is of a still heavily wooded countryside before this period which seems to have experienced drastic deforestation over a fairly short interval of time. Such evidence is not inconsistent with the picture of sudden and widespread conflagration described by Gildas.

There is further physical and documentary evidence, also, of a sharp deterioration in global climate in the aftermath of these events. The northern treeline in Canada and Scandinavia receded by a hundred miles or so (figure 2), whilst the Annales Cambriae speak of days as dark as night in AD 444 or AD 447, consistent also with Gildas's reported failure in the production of wine. In the history of Northern Britain, too, recounted much later by Nennius, there is mention that Germanus made another visit after the disaster, but there are curious and unexplained references to the destruction of Vortigern's fortress 'by fire sent from heaven [as] the fire of heaven burned'.

Other literary sources of less certain pedigree speak of Vortigern, who had previously consulted his magi (astronomers) and withdrawn to his fortress, actually perishing during this incident. We would not then expect the official delegation from Gaul to have any subsequent contact with Vortigern and indeed it is reported that Germanus was received by a certain Elaphius who was said to be 'the first man of the district'. The latter, in view of the destruction of central authority in the land and of the destitute state of the population and countryside, might well under these circumstances have been persuaded as to the merits of a mass exodus across the Channel to relieve the immediate hardship in Britain whilst also providing much needed reinforcement for Roman Gaul against the recurring threats from Visigoths and Huns. But as the exodus went ahead, it seems that the former allies and mercenaries of Anglo-Saxon origin were reappearing in Britain to claim settlement rights that were said to have been agreed previously with Vortigern. Such hints as there may be in this course of events of a heaven-sent force which effectively ruined Britain, argue in favour of a celestial rather than an Anglo-Saxon agency for 'the fire of righteous vengeance' cited by Gildas. Indeed the 'battering rams' which he also cited, in their Latin equivalent, are 'aries', a term with obvious connotations relating to stars as much as sheep (see also chapter 1), even to the extent of a specific association with the largest known cosmic swarm (chapter 10). Commentators in China and Spain, it is worth noting, also record an unusual observation of a 'strange comet' in 441 AD.

Subsequent reports on the continent of a sequence of events leading to a disaster in Britain before 442 are in fact now seen as making quite exaggerated claims for successful invasion by the Anglo-Saxons at this time. However they may justifiably be seen as evidence that piratical opportunists were seeking to establish footholds in Britain. Thus, although the Anglo-Saxon dominance of Britain from this time was once accepted as fact, the evidence can now be understood as pointing to a cosmic catastrophe unrelated to continental invaders and subsequently, during the latter half of the fifth century and the first half of the sixth, of a Celtic Britain which was restored to something of its former strength by members of the nobility, returned from Gaul (cf. Table 2).

Indeed the main Anglo-Saxon invasion taking over Britain is now seen as having occurred at least a century later than the catastrophe, at a time which corresponds to the earliest known returns, according to the calculations of the German astronomer Klinkerfues,[62] of intense meteor showers due to Comet Biela (chapter 9). The tendency in Britain five centuries later, when Anglo-Saxon overlords were replaced by a new generation of nobles from northern Gaul, to hanker after a glorious Arthurain age in which an ancient celestial figure in the form of Merlin also commanded great respect, does then become more comprehensible as evidence of a presumed, also greatly feared, celestial influence during the Dark Age.[63] Indeed a Dark Age whose aspiring kings were much given to heading their genealogies with celestial gods, in general accordance with a still widely accepted principle of divine election (cf. plate 18), is likewise the more comprehensible. It is an interesting question, in fact, whether the whole of the first millennium AD should be understood in terms of a crusading ethic amongst the divinely elected, inspired by celestial events, first setting Anglo-Saxons inter alia on the move, only to be replaced by the Vikings in due course.

Later, the exploits of the Vikings were on a grandiose scale, embracing the whole of Europe.[64] But they grew out of a Dark Age already centuries long and our knowledge of their origins is very sparse. The best documented of the Viking raids, and also the earliest recorded in the west, is the plundering and destruction in 793 of the church and monastery on the tiny defenceless island of Lindisfarne. The Anglo-Saxon Chronicle states that terrifying omens – lightning and flying dragons – which were witnessed that year were followed by famine, and that shortly after these dolorous events, in June, heathens fell upon the island community and pillaged God's house there. During the restorations of the twelfth-century priory on Lindisfarne, archaeologists found a curious carved stone apparently dating from soon after the assault on the earlier monastery and illustrating the dreadful event. On one side of the stone were carved the various cosmic symbols, the Cross, the Sun and the Moon, God's hands, and worshippers at prayer. On the other side the attackers are seen, swinging their swords and battle-axes as they advance in single file, dressed outlandishly in thick jerkins and narrow trousers: invaders perhaps who were preceded by 'the full flame'? The Vikings consciously saw themselves apparently as marauders who pounced from concealed positions, as if from creeks

(a)

(b)

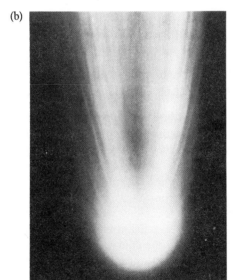

(c)

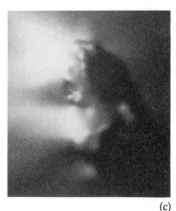

Plate 9. Perspectives on Halley's Comet. (a) As represented on the Bayeux tapestry, during its 1066 apparition (Tapisserie de Bayeux et avec autorisation spéciale de la Ville de Bayeux). (b) Crowned with great horns? As photographed from Mt Wilson Observatory during its 1910 apparition. (c) The nucleus during its 1986 apparition, photographed by the European Giotto spacecraft (by kind permission of Dr H. U. Keller, Max Planck Institut für Aeronomie, Lindau/Harz, FRG. © MPAE, 1986). The nucleus is an extremely black, rotating object with dimensions about 16 × 10 × 9 kilometres. The tail of the comet is made up of gas and dust ejected in jets from a few active areas on the surface of the nucleus. During a close encounter with a brilliant comet, individual jets would be seen, creating a spiral or possibly even a swastika in the sky (cf. plate 19).

or caves, supposing the etymology of the word is understood, but to Spaniards they were 'heathen wizards', to Germans they were 'ashmen'. There are strange associations here that have not been explained. Historians indeed have not even remarked on the dramatic increase in the flux of meteor storms and fireballs during this period, although we have good records of events in the sky from the Chinese astrologers.

This renewed activity in the sky evidently reached its maximum during the eleventh century AD, the peak coinciding, by chance, with the conspicuous reappearance of a bright comet in 1066 (see plate 9). Inevitably, warrior kings, builders of vast keeps in the style of an earlier Mycenean age, would relate their rise to power to such signs from heaven whilst the Christian Church would draw upon the resources of the people in raising huge cathedrals and other places of worship, still modelled paradoxically on the pagan cosmos, in preparation for the 'millennium'.

At this time too, it is reported quite specifically by near-contemporary historians in the 'Recueil des historiens des Croisades' that Pope Urban II convened the Council of Clermont in November 1095 for the purposes of authorizing the First Crusade, with all the far-reaching consequences that it set in train, immediately following a violent meteor storm[65]:

In the time of the emperor Henry IV . . . according to the prophecies in the Gospels, everywhere nation arose against nation and kingdom against kingdom; and there were great earthquakes in divers places, and pestilences and famines and terrors from heaven and great signs. And because already in all nations the evangelical trumpet was sounding the coming of the Last Judge, the universal Church beheld throughout the whole world the portents in prophetic signs. . . . [Thus] when it was God's will and pleasure to free the Holy Sepulchre [at Jerusalem], in which his son had lain for the sins of men, from the power of the pagans and to open the way to Christians desiring to travel there for the redemption of their souls, he showed many signs, powers, prodigies and portents to sharpen the minds of Christians so that they should want to hurry there. For the stars in the sky were seen throughout the whole world to fall towards the earth, crowded together and dense, like hail, or snowflakes. A short while later a fiery way appeared in the heavens; and then after another short period half the sky turned the colour of blood.

The resilience of crusading Christian theology was thus once again tested, and seemed to ride high on the renewed fears of Doomsday and world-end.[66] Indeed, during the centuries that followed this meteoric spasm, the Christian Church maintained its power, but as the sky ceased once again to be important, the Church moved to embrace Aristotelian principles rather more readily than the feudal warlords who presumably still depended on signs from heaven to underpin their kingships and influence. Thus in the wake of the Arthurian age and the decline of Merlin, for such the millennial meteoric spasm appears to have been, the Christian Church centred on Rome enjoyed its period of maximum influence. But in due course, yet again, in the fourteenth century, strange portents in the sky (chapter 1) and an inclement climate seem to have re-emerged to add to the plight of mortals on earth. It is no surprise that, in the respite that followed – the European Renaissance so-called – the authority of the Church and the fundamental nature of the cosmos should once again be questioned.

8

Felony Compounded

THE road from Ptolemy to Copernicus was thirteen hundred years long, and in all that time scarcely a new idea found its way into astronomy. The Academy at Athens was closed whilst that at Alexandria was stultified by ideology, the Great Library eventually being burned to the ground, either by Christian fanatics or Arab conquerors. The goal of unifying physics and astronomy, already something of a charade in Ptolemy's time, had become a remote prospect. However, following centuries of which it could be said 'All curiosity is at an end after Jesus, all research after the Gospel. Let us have Faith, and wish for nothing more', the opportunity to rethink seems again to have been at hand.

Remnants of Greek science eventually found their way to the new medieval universities of Europe by way of the Arab kingdoms of Spain and Sicily (plate 10), and something of the way in which the Greeks first came to study the planets seems to have been reconstructed. More significantly perhaps, through institutions provided by the medieval Church, a tradition of scholarly teaching and study came to be revived. By the middle of the thirteenth century, indeed, Thomas Aquinas was arranging that the Aristotelian system was once again theoretically wedded to Christian theology, so ensuring that any future reconstruction of physics outside Aristotelian or Ptolemaic lines would not be a task for the faint-hearted.

Despite this, there were aspects of Aristotelian cosmology that continued to worry the purists; particularly the central position of the Earth and the location and status of comets. For example in the fifteenth century the German cardinal Nicolas of Cusa, in his treatise *On Learned Ignorance*, disputed the Aristotelian idea that space could be bounded by an outermost sphere; and if there was no boundary to space, how could there be a centre? Earth, Sun or Moon could be imagined at the centre of the universe arbitrarily. And if the

Plate 10. A Muslim astronomer observing a comet with a quadrant (Topkapi Sarayi Museum Library).

universe had no centre to which motions could be referred, why should the Earth not be considered in motion?

Such questions had not been properly settled and around the middle of the second millennium AD, Copernicus (1473-1543) indicated that perhaps the Sun should be placed in the middle. The concept of a central Sun was at first deflected by the Church: thus an arrangement by one Osiander ensured that Copernicus' work was prefaced with an unsigned disclaimer stating that the new idea was merely a mathematical device for the purposes of easy computation, and not the physical truth. The Church, Roman and Protestant, had little difficulty therefore in suppressing this new idea and later attempts to revive it led to Bruno (1548-1600), in the hands of the Inquisition, being burned at the stake and Galileo (1564-1642) being placed under house-arrest. However, within a century of the death of Copernicus, the simplicity of the new system as against the complexity of the old was beginning to tell.[67]

It was the careful observations of the Danish astronomer Tycho Brahe which eventually rendered necessary a radical change. The ground had been prepared by the appearance, in 1572, of a brilliant star in the sky (known now to have been a supernova), which was hard to reconcile with the perfect, unchanging sky of Aristotle. A great comet then appeared in 1577 and Tycho, by careful observations, was able to show that even at its nearest approach to Earth, the comet was still well beyond the Moon, contrary to the Aristotelian view that comets were mere atmospheric phenomena. It was still conceivable, of course, that comets were indicators of divine intention, and indeed the appearance of this comet unleashed much sermonizing and writing of blood-curdling tracts, but those of the time who were less eschatologically inclined were now able to take a more prosaic view.[68] Kepler, for example, simply assumed comets were sweeping past the Solar System in straight lines while Galileo treated them as mere optical illusions!

The latter was not a particularly tenable point of view on any account and we find Newton (1642-1726), somewhat later, preferring the Keplerian assumption: not out of conservatism, as it happens, but because he wished to avoid comet encounters doing any damage to the Solar System as he perceived it. Thus Newton (plate 11), like Aristotle before him, considered the Solar System to be a divine creation, set to run like a clockwork machine; and stray comets passing through obviously put the system very much at risk.[69]

Engraved by W.T.Fry.

SIR ISAAC NEWTON.

OB.1727.

FROM THE ORIGINAL OF KNELLER, IN THE COLLECTION OF

THE RIGHT HON^ble THE EARL OF EGREMONT.

Plate II. Sir Isaac Newton (Mansell Collection).

In 1680, however, another important comet was observed. Its path across the sky was accurately tracked by the Astronomer Royal, Flamsteed, at Greenwich. The orbit of this comet was shown to be, not a straight line, but a parabola coming within the planetary system. It says something for Newton's understanding of gravity at this stage that he did not treat the parabola as an illustration of his gravitational law. In fact, he at first refused to believe Flamsteed's observation and only later came to regard the orbits of comets as another aspect of the divinely ordained law of gravity, and the latter as universal. But with a subtle change of emphasis, Newton now insisted that encounters of comets with planets should be seen as providential rather than catastrophic events, the means whereby the planets were occasionally revitalized. Very little attention has been paid to this interesting circumvention of Newton's but its impact on science has in some respects been almost as significant as that of his law of gravity. Let us follow the matter a little further.

Newton was, of course, not alone in reflecting on these questions and by this time the possibility of a disaster through cometary impact on the Earth had become a matter for general debate. That the Earth's axis had shifted in the past due to the close approach of a comet was accepted as a serious proposition. A future encounter, it was thought, might bring the world to an end through overheating it. Scholars seriously discussed whether the conflagration of Phaethon and the flood of Deucalion or Ogyges had been caused by a celestial body. The flood of Ogyges was noted for example as having been attributed to the arrival of a comet Typhon, and there was also a Biblical connection for Roman historians had declared the plagues of Egypt to be a contemporary event. Such issues as whether the floods of Noah, Ogyges and Deucalion were one and the same were also raised, and the outcome of all these enquiries was a growing interest in terrestrial catastrophism and ancient chronology. On the one hand, it seemed that historical studies might help to determine the periods of recurring comets, and on the other, it seemed that periodic returns might be used as a celestial clock to plot the course of major historical events, the epochs for example of Abraham, Moses and Christ.

Then, by chance, another comet appeared in 1682. Halley was one of the first to note that its orbit was that of the comets of 1531 and 1607: a celestial clock had been found. This particular comet was to take Halley's name in due course (following its predicted return in

1758) but Halley's more immediate interest was in deriving a period for the 1680 comet. A value of 575 years was obtained, incorrect as it happens, which William Whiston (Newton's successor in the chair of mathematics at Cambridge) made use of in calculating earlier apparitions. This quite fictitious orbital period had the comet returning at the time of Caesar's death in 44 BC, and an earlier return was supposed also to coincide with Noah's flood in 2342 BC. Whiston's book describing these researches, dedicated to Newton, rapidly came to be seen by clerics and natural philosophers alike as a major step forward in the advancement of knowledge. Seemingly for the first time, observational science and received Biblical knowledge were in excellent accord. It was a history of catastrophe moreover, and the day of judgement could even be at hand.

Whiston did not rest on his laurels, however. He thought the course of events simply required a more matter-of-fact interpretation of the Bible, and he tended to play down the role played by Christ. Thus, although Jesus's standing as a prophet could be secure, his divine attributes seemed again open to question. This was too much for most of Whiston's contemporaries, who saw this as an attempted revival of the heretical doctrine known as Arianism, and the clerical establishment soon sought to distance itself from Whiston. In the meantime Newton, boxed in by these new implications of comets, took it upon himself to emphasize again what he now claimed to be the fundamental role of comets, namely their ability to deposit new material on to the stars and planets and their potential for doing good to the Earth. The calm voice of reason was speaking. Scaremongering was naturally added to the list of Whiston's faults and it was not long before he was dismissed from his post. The message was clear: the Earth was not to be disturbed.

There is evidence that Newton was in fact well aware of the catastrophic potential of comet impacts. These did not, however, fit easily into his clockwork, predetermined universe, and he never publicly raised the issue. Whether we should accuse Newton of dissembling over the issue is perhaps uncertain but there is no question that many who followed Newton, in England especially, were soon eagerly endorsing the harmless role of comets, now that they were seen to obey mechanical laws and could no longer be regarded as signs from Heaven.

The return of Halley's Comet in due course, in accord with

gravitational law, merely served to reinforce the new paradigm – as also did Lexell's comet some years later. This comet passed the Earth without exerting any noticeable change on the planet's orbit. This was good evidence that comets were very much smaller than planets, thereby diminishing still further any lingering concern over their menace. By the end of the eighteenth century, then, Newton's public view of comets had gained considerably in strength and we find William Herschel, George III's astronomer, putting forward the opinion that comets were to be taken as interstellar objects purposely weaving their way among the stars including the Sun, with the specific intention of replenishing their fuel, and if necessary, planetary life as well. Newton's theory had therefore developed rather successfully into an all-embracing cosmological view of some permanence and continuity.

The new discoveries of meteorites and asteroids at the beginning of the nineteenth century might have disturbed the *status quo* by raising the spectre of celestial hazard. In the early decades of the nineteenth century also, Comet Biela was seen to break up and intense meteor showers appeared in 1832 and 1833. The latter shower, seen in North America, led to surprise, panic, terror and a widespread belief that the Day of Judgement was near.[70] There was at this time also a revival of catastrophist views especially among scientists of the French Academy.

However, the revival was short-lived. The generally small sizes of the meteorites and the apparent confinement of the asteroids to a belt between the orbits of Mars and Jupiter must have allayed any fears. Further, by the middle of the nineteenth century, there was a growing realization that many comets in short-period orbits rapidly decayed under the influence of heat from the Sun into harmless streams of dust. Indeed it seemed that Nature was able to cooperate with Man by turning comets into dust which fed into the atmosphere to be witnessed always as shooting stars. Comets were thus harmless again and there was even a growing appreciation that they were probably not interstellar, as Herschel had supposed, but had always belonged to the Solar System.

The Universe was thus beginning to seem again a rather cosy affair, the Sun pre-eminent in space and the Earth on a sort of eternal merry-go-round, with the good fortune to be entertained from time to time by firework displays against the background of more permanent stars. By the middle of the last century, therefore,

Newton's long-established theory of comets appeared to have been vindicated: celestial portents had been well and truly tamed, and the benevolent cosmos of Aristotle and Paul had been reconstructed.

Within about a hundred years of the period which we now call the Enlightenment, a situation had emerged in which simple mathematical laws (those of motion and gravity) appeared to govern absolutely the forces of Nature. Such a situation was unprecedented and was the basis of the singular confidence of nineteenth-century physical scientists and Victorian society generally. Admittedly, if one were to examine the details, there were difficulties with the mechanical aether which was supposed to transmit waves of light as well as the forces of gravity, but the solution of these particular difficulties was widely assumed to be just a matter of patience and time. The mechanisms involved were, it was thought, likely to be simple and could fall easily into place once there was a proper grasp of all the experimental phenomena. With our knowledge today of all the surprises yet in store and of all the great upheavals in physics that were to come,[71] it is difficult perhaps to grasp the real feeling of mastery over the forces of Nature that prevailed in the nineteenth century. This feeling also extended to the astronomical environment in view of the latter's presumed harmless character arising from Newton's theories. Thus, while gravity could be seen to pervade the universe and provide a smooth undercurrent of physical control over the behaviour of matter, it was also possible so far as comets were concerned to draw a clear demarcation between the unharmed Earth on the one hand and the non-interventionist cosmos on the other. The impact of this revitalized Aristotelian arrangement was far greater than is often realized these days; for never before had there been created such an impression of security in relation to the astronomical surroundings. The new benevolent cosmos underwritten by Newtonian laws could even now be interpreted as the ultimate triumph of a tranquil Christian tradition over fiercer Judaic ones and those of presumed less-enlightened faiths.

There are tempting parallels here between the British Empire at its zenith and its Hellenic predecessor. In both cases the world was, in some sense, under control. Certainly Victorian society had its chartist movement, an unpopular monarchy, great inequality of wealth and so on, whilst the Hellenic kingdoms were frequently at war with each other; nevertheless in both cases a sense of confident control over destiny seems to have pervaded these societies. And in both cases the universe was understood as a more or less non-interventionist backdrop to the affairs of the Earth. It may be no accident that, like its Hellenic predecessor, an empire at its zenith should have provided the setting within which the universe was to be so understood, and no surprise that earth scientists and biologists now felt free to explain terrestrial and biological evolution unhindered by any thought of external interference.

In like manner, historians and sociologists began to look to purely human factors dominating the course of historical events. The belief can also be found in classical times, being held for example by the ancient historian Thucydides. But now a sharp division of interest emerged between those who explored the material world and those who explored the human one, the latter secure in the knowledge that, apparently, any external influence had finally been removed. This division into 'two cultures' evidently has no valid foundation and there is some awareness that it causes substantial damage in the educational and economic spheres at least. What is not understood is that in addition, by supposing the affairs of Earth to take place in splendid isolation, in an irrelevant cosmos, the massive hazards which mankind may face from the astronomical environment have in effect been programmed out of public consciousness.

Of course, this argument is a gross simplification and a partial caricature, but it is possible to wonder nevertheless whether our distorted view of history today has not unleashed on western civilization such a sense of superiority that we spurn the sky and put our future in peril. Certainly we cannot escape the fact that the loss of a cosmic perspective arising out of Newton's fundamental thinking has now penetrated more or less every branch of learning to the extent that it goes largely unrecognized. For example, there is a widespread and mistaken view nowadays, especially among earth scientists and Biblical fundamentalists, that the issue of catastrophism in Earth history was decided at this epoch, the middle of the nineteenth century, by geologists and biologists debating the forces of natural selection and the nature of geological change. In fact, the uniformitarian framework had already been agreed by astronomers, as we have seen, and it had simply become a matter of convincing oneself that contemporary terrestrial processes were enough to explain evolutionary change, as revealed for example by the fossil record. Admittedly catastrophists like Buckland and Cuvier opposed the trend but the debate was short-lived. As it happens, the geologist Lyell and the natural historian Darwin, and their followers, took the lead over this issue because they were not particularly bound by any Biblical or physical conventions regarding the age of the Earth; but insofar as they needed support for the slow action of virtually undetectable forces, the physical scene had already been set.

Over the century or so following the publication of Darwin's theory of evolution, catastrophism lost all respectability. From time to time, attempts were made to revive what might be called 'Biblical' catastrophism, the idea that astronomical disasters have occurred on Earth in historical times. One such attempt was published in 1823 by the German linguist Radlof (plate 12). This was 127 years after Whiston's *New Theory of the Earth*, and 36 years before Darwin's *Origin of Species*. At the time Radlof wrote, only four asteroids had been discovered in the large gap between the orbits of Mars and Jupiter, but it was widely thought that they were the remnants of a planet which had exploded and that numerous fragments would eventually be found. Radlof developed a hypothesis which was both consistent with the scientific knowledge of the day and also seemed to account for the ancient tales of catastrophe. According to him, a large planet in orbit between Mars and Jupiter was shattered by collision with a comet. One fragment, Phaethon, struck the Earth, giving rise to tales of celestial battle in which the monster Typhoeus was hurled to Earth. Another fragment became the planet Venus, undergoing close encounters with Mars on its way in to its present orbit.

Sixty years later, world end traditions were again similarly interpreted by the Irish American politician Donnelly. His book *Ragnarok: the Age of Fire and Gravel*, published in 1883, created a considerable stir when it came out. In it, Donnelly argued that the Earth had had a close encounter with a comet, which had left its mark in the form of extensive gravel and till deposits in America and Europe.

The most famous (or notorious) of the 'Biblical' catastrophists, again in America, was Velikovsky, the author of *Worlds in Collision*, who in 1950, from his studies of ancient tales, revived the old notion that a comet had shifted the Earth's axis, and even resurrected Radlof's notion that the planet Venus had hurtled past the Earth around 1500 BC, flinging Mars around in the process. There were

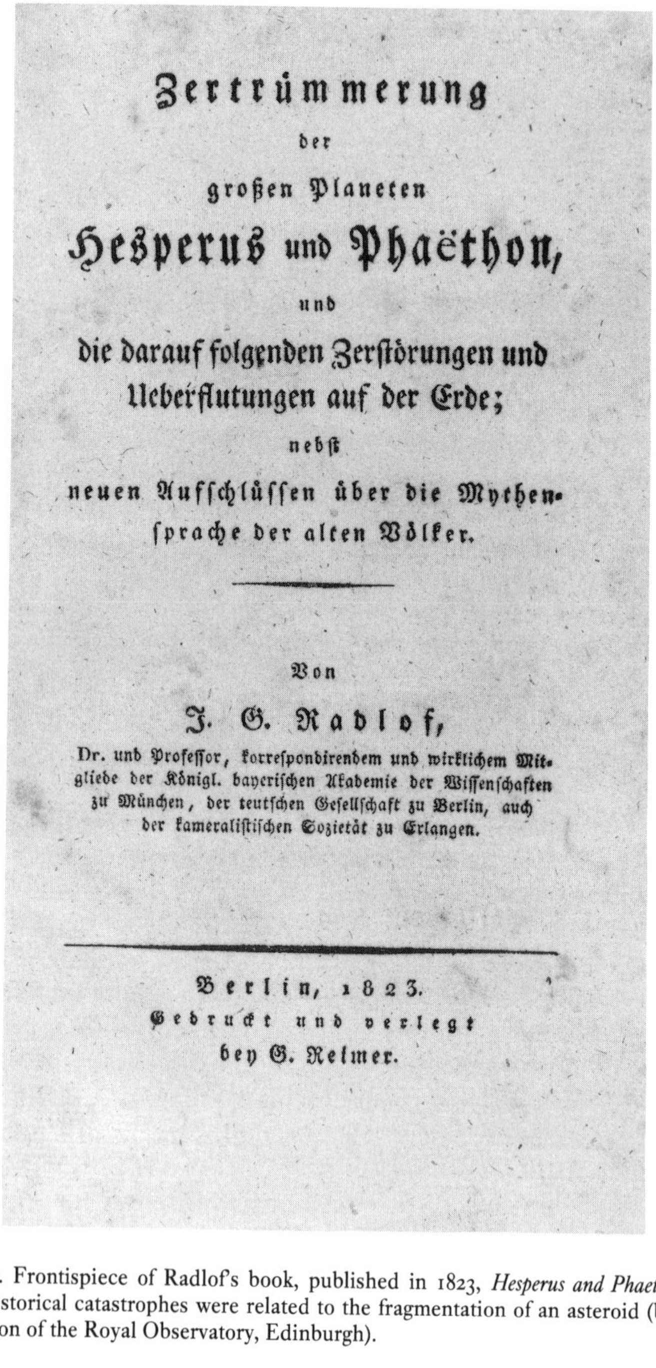

Zertrümmerung

der

großen Planeten

Hesperus und Phaëthon,

und

die darauf folgenden Zerstörungen und
Ueberflutungen auf der Erde;

nebst

neuen Aufschlüssen über die Mythen-
sprache der alten Völker.

Von

J. G. Radlof,

Dr. und Professor, korrespondirendem und wirklichem Mit-
gliede der Königl. bayerischen Akademie der Wissenschaften
zu München, der teutschen Gesellschaft zu Berlin, auch
der kameralistischen Sozietät zu Erlangen.

Berlin, 1823.
Gedruckt und verlegt
bey G. Reimer.

Plate 12. Frontispiece of Radlof's book, published in 1823, *Hesperus and Phaethon*, in which historical catastrophes were related to the fragmentation of an asteroid (by kind permission of the Royal Observatory, Edinburgh).

differences of detail: whereas Radlof had Venus as a planetary fragment flung out of the asteroid belt, Velikovsky had it thrown out of Jupiter. Time, however, had marched on: what was perhaps seen as tenable in 1823 was now patently absurd in 1950. The decline of catastrophism, as a serious scientific proposition, was complete.

It is interesting to note that the Biblical catastrophists were all 'outsiders' of one sort or another. Radlof was at first regarded by his contemporaries as a brilliant linguist, but as his ideas departed more and more from the mainstream so he became more and more isolated, and he died in obscurity, even the date of his death being unknown. Donnelly was regarded as a talented orator and statesman, but he seemed unable to toe any party line for long and his increasingly maverick opinions ensured that his political career was ultimately a failure. Velikovsky was a psychiatrist who had studied under Freud, but whose incursion into the domain of astronomers was to prove an intellectual disaster.

What is significant in the Velikovsky affair is not so much his scientific proposition, which is an easy target, as the vitriolic reaction of the American astronomical community to it. Little attempt was made by the critics to evaluate the historical information; that there might be more things in heaven and earth than was apparent in the astronomy of 1950, that Velikovsky and his predecessors might however dimly have been detecting the signal of real events in the noise of ancient myths, does not seem to have occurred to them: celestial portents had been more than just tamed, they had become unthinkable.

We have now traced the labyrinthine story of mankind's reactions to the events in the sky. We have seen how preoccupation with the sky, and fear of it, was an integral part of the earliest civilizations. There was a clear perception that catastrophe might from time to time be visited on the Earth from above. But as the gods faded from the zodiac there was, inevitably, a shift of perception. A downgrading of comets as celestial portents was both politically opportune, and philosophically necessary if the theory of a perfect sky was to hold.

Of course, the linkages between religion, government and philosophy are part of the complex stuff of history, and our working out of the above view in earlier chapters is inevitably simplistic. But if there is a kernel of truth in the proposition that environments interact – natural, intellectual and political – then it cannot be avoided that exciting performances from the sky, occasionally associated with something like a nuclear impact, will generate a particular view of the world, and that the decay of these performances will leave that world-view high and dry. Likewise, if there is a kernel of truth in the proposition that meteors were once frequently recognized demons, with a not wholly unpredictable role to play in the incidence of weather and disease as well as in the survival of 'chosen people', then, nor can it be avoided that a particular, seemingly plausible, view of the world will in due course be seen as dealing only in natural magic and occult virtues. The surprise, of course, is that there may be more realism in the hidden forces of medieval magic than there may be in some of the physical concepts now advanced within a modern scientific rationale.

We have seen how an ancient belief in swarms of wandering bodies that would sometimes emerge to strike the Earth gave way in post-Aristotelian times to the nonsensical idea that planetary conjunctions were of significance for life on Earth. Modern news-papers, presumably guided by editors with a preponderantly classi-cal education, sublimely unaware of both earlier and later developments, still persist in presenting this perverse form of astro-logy, involving horoscopes, to the twentieth-century public. We have seen how in the eyes of Christian apologists determined to root out apocalypse, comets ceased to be material objects and became celestial signs free for the expert to interpret. Several times over the last two millennia, phenomena from the sky were seen to threaten world-end: once at the beginning of the present era (200 BC–AD 100), again during the first millennium (AD 350–450, AD 550–650, AD 800–1100) and then again, albeit less impressively, perhaps three times during the present one (around AD 1350–1450, AD 1650 and the early decades of the last century). Several times, the apparent threat lingered for centuries, paradoxically adding weight to the Christian myth. Then, towards the end of the medieval period, as the Christian logic appeared to falter, there was an opportunity to reassess the question of the interaction between the Earth and its environment. But the Christian grip was strong and Newton, like Aristotle, had his subtle vision of perfection which carried great influence. He claimed not to 'deal in conjecture'; and the way forward, perceived as such to the present day, was to build on the secure mathematical foundation of his laws. We have seen how his revelations of cosmic security and cosmic law were to be the lynch-pins around which twentieth-century science revolved.

But now, the space age has arrived, riding on scientific laws, and the importance of past impact cratering on planetary surfaces has been recognized. The extent of the asteroidal population is also now being revealed. These and other discoveries are now being put together as cause and effect. Most of the evidence, as we shall see, indicates that asteroids are probably defunct cometary material and that their supply into the central region of the Solar System is maintained by the intermittent flux of giant comets from a comet cloud surrounding the planetary system.[72] The implication is that swarms of asteroids periodically exist in Earth-crossing orbits and that these are responsible for producing an erratic sequence of cosmic winters, sudden coolings of the globe.

There have in fact probably been at least two successful and several more aborted cosmic winters during the last five thousand years. The strength and epoch of the next such global cooling cannot be calculated at the moment because the celestial body or bodies likely to cause it has (or have) not yet been detected by astronomers. In fact there is little interest in a search, although the effects of such an encounter may be broadly as horrific as those of a nuclear war. As always, the unthinkable becomes the excuse for not looking.

Some lemmings, a story has it, rush over cliffs. Others, with the successful breeding of two thousand years, just sit on a globe and wait . . .

II

THE BULL OF HEAVEN

9

Celestial Mechanics

To the casual observer on a clear dark night, away from city lights, the stars appear as points of light on a huge dark dome. Stretching across this celestial vault, there is a broad, irregular band of light known as the Milky Way, a river or path over which, in earliest times, were said to have trodden ghosts, pilgrims, souls and many other creatures of the mind. Binoculars reveal the Milky Way to be made up of myriads of stars. The Sun is one such star and it would only have to be moved ten or twenty light years away to be completely lost against the background of the Milky Way. In fact, contrary to appearances, the latter is a mighty disc some 100,000 light years across, containing some one hundred billion stars which are on average just a few light years apart. The whole system is now known as the Galaxy and it happens that the Sun is immersed within it close to the central plane and some 25 light millennia from its central nucleus.

To get a better idea of what these huge distances mean, we can scale down the dimensions of the galactic disc to those of a large city such as Greater London; its thickness is then about 200 metres and the average distance between stars is about a foot. On this scale, the planetary system is only about a third of a millimetre across; and we can readily see that the Galaxy is largely empty space. If the age of the Galaxy, which is about ten thousand million years, is similarly compressed into one year, the Sun and its planets are less than 5 months old whilst the dramas of the present ice age have been acted out only during the last 15 seconds of the year. To some degree then, our environment is simply a matter of perspective: the astronomer tends to see the Earth as a tiny vulnerable speck drifting with the Sun in a huge galactic wilderness; the geologist tends to see it as a massive self-contained globe evolving under the influence of strong internal forces; whilst the historian tends to see it as an inert platform, without relevance to the performances that are staged on it.

The vast astronomical distances tend to give one a misleading impression of the activity in the Universe however. The stars and planets are in fact in a continual state of motion under the influence of gravitational fields but in the case of stars, the distances are so great that very little is perceptible to the naked eye. Planets on the other hand are much closer and appear correspondingly more active, their orbital periods being measured in years. Now, a planet orbiting a solitary Sun would continue in an unchanging elliptical orbit forever. It is not so obvious however whether this simple state of affairs would continue in the real world. Thus, while planets disturb each other through their mutual gravitational attractions, and there are small deviations from perfect ellipticity, the cumulative effect of these changes is known to be extremely small over periods of several million years, so the stability of the Solar System planets is mathematically assured over these timescales. Beyond that, opinion about the stability of planetary orbits is based on physical rather than mathematical insight. For example it has been calculated[73] that, if the Earth were to move closer to the Sun by about 5 per cent, a 'runaway greenhouse' effect would set in, generating an unstable rise in the Earth's temperature to the extent that the oceans would boil. On the other hand, a 10 per cent movement in the other direction would freeze them solid. In either case it is probably reasonable to suppose that life on Earth would come to an end. These are tight tolerances and suggest that the Earth's orbit has not changed very much over the last billion years or so. It is very likely also that a large change in the shape of the Earth's orbit would give extremes of climate probably beyond the ability of life forms to survive. These are reassuring arguments: one does not need to worry about whether the Earth might be flung into the Sun or into interstellar space, or whether it might find itself on a collision course with Venus. At first sight then astronomers are in a position to agree with geologists and historians regarding the seemingly harmless nature of the astronomical environment. Everything indeed would be fine were it not for a single discordant fact – the existence of comets.

The outermost planet of the Solar System, so far as we know, is Pluto, a small body at a mean distance of four light hours from the Sun (although it is currently inside the orbit of Neptune). Out beyond the known planets, we pass through uncharted territory: there could be comets, asteroids, planets or moons at distances of light days or light weeks, but we know nothing of them. Eventually, light months to a light year beyond the Sun, we pass through a huge cloud of mostly kilometre-sized bodies in a deeply frozen state: these are the comets. There are at least as many comets orbiting the Sun as there are stars in the Galaxy. Typically, a comet orbits the Sun in 3 to 6 million years, as against 250 years for Pluto, twelve for Jupiter and one for the Earth. Comets are made of ice and dust, and possibly larger bodies, and they are composed of the specific 'universal mix' of chemical elements that is found in stars, less hydrogen and helium, which remain gaseous even at a few degrees above absolute zero. And it is here, on the icy outer margins of the Solar System, that we find the potential for chaos on Earth. The existence of this cloud of comets was only established to the satisfaction of most astronomers around 1950.

So long as the stars were seen as the only significant component of the Galaxy, the comet cloud could be regarded as relatively secure and harmless. Every million years or so, a star would drift through the cloud and gently churn up the orbits of the comets, only those whose orbits led them almost into the path of the intruder being disturbed sufficiently to be thrown off course, possibly into interstellar space; the overall effect of these stellar encounters is weak enough to leave the bulk of the comet cloud firmly intact. This quiescent picture was almost universally held until the late 1970s, and is entirely in keeping with a view of the Earth as a body isolated from its environment and undisturbed by extraterrestrial visitation. In recent years however, with the discovery by radio astronomers of cold, dark massive nebulae in the Milky Way known as molecular clouds, it has come to be realized that the reality is different, that the comet cloud is in fact transient and unstable, and that the Earth is very much affected by its astronomical environment.

The patchy irregular structure of the Milky Way is obvious on any clear night and has long been known to be due to dark clouds of gas and dust obscuring the starlight behind them. However nobody knew how massive these nebulae were likely to be and it is only through the studies by radio astronomers that we have now become aware they are in fact a major constituent of the galactic disc, and include the most massive single bodies in the Galaxy. This fact had been missed because the clouds are for the most part extremely cold, only a few degrees above absolute zero; and at such low temperatures hydrogen, the commonest element, exists largely in molecular form which is undetectable in ordinary light. However the presence of molecular hydrogen can now be inferred through radio emissions from carbon monoxide which acts as a tracer for hydrogen in concentrations of cold cosmic gas. It turns out that a typical giant molecular cloud is about 100 light years across – 30 metres on the Greater London scale – and that they often contain dense concentrations of young stars and, in all probability, enormous numbers of newly formed comets as well, circulating freely within the nebula. The mass of a giant molecular cloud may be half a million times that of the Sun. A few thousand of these monsters orbit the Galaxy, confined within the flat plane of the Milky Way, and the Sun has probably penetrated ten or twenty of them in the course of its history, and had close encounters with many more.

The effect of such encounters is calculable: the Solar System must have been stripped of its comet cloud not once but many times in the course of its history. Nevertheless it is there! It follows that the cloud must from time to time be replenished. The source of the replenishing comets, and how they come to be emplaced in orbits a light year from the Sun, are interesting unresolved issues at the present time; indeed the ultimate origin of comets has been a controversial question for hundreds of years.[74] But this need not concern us here: the important thing is that we have now realized there is a *cosmic* process with the potential for *terrestrial* catastrophe. For during the Sun's bumpy passage near or through a molecular cloud, the comet cloud is subjected to severe gravitational buffeting. The result is that, not only are many comets flung into interstellar space, others are thrown into orbits which take them, several million years later, directly into the planetary system.

It has also been realized recently that over and above the effects of molecular clouds, which are encountered irregularly, the gravitational influence of even larger structures in the Galaxy is also important. These larger structures, spiral arms (see plate 13) and even the galactic disc itself, create tides. At the surface of the Earth their strength is minute, but at the distances characteristic of the comets they have a profound effect on cometary orbits. As the Sun goes round the Galaxy, passing in and out of spiral arms and up and down through the galactic plane, the strength of the galactic tide ebbs and flows in a periodic manner. Periods of high and low tidal stress on the comet cloud occur, leading to periods of high and low comet flux into the planetary system.

There are therefore two main sources of disturbance of comet

Plate 13. A typical galaxy, showing spiral arms: Messier 83, a large, almost face-on spiral galaxy perhaps 10 million light years away in the constellation Centaurus. This galaxy probably resembles our own. It contains strong, complex spiral arms with narrow dust lanes, young hot stars and cold, probably comet-bearing nebulae. An exotic radio source lies at the nucleus of this galaxy (photograph reproduced by kind permission of the Royal Observatory, Edinburgh).

orbits, one erratic and one periodic. The variable galactic tide gives periods of high and low comet influx depending on the Sun's position in the Galaxy, and superimposed on this regularity, molecular cloud passages give rise to showers of comets into the terrestrial neighbourhood, each shower lasting for something like three or four million years. By ordinary standards, the term 'shower' is a trifle exaggerated since the flux implies only a few comets per annum in the vicinity of the Earth's orbit; nevertheless it is entirely appropriate on the cosmic scale. It will turn out that the Sun's current position on this grand, galactic scale is very relevant to our assessment of the celestial hazards we face at the present time.

The fact that the flux of comets varies does not of course change the nature of comets; all other things being equal, the assumption that comets are harmless might still apply. And indeed, despite their occasional spectacular appearance due to the release of volatile gases and dust close to the Sun, with one or more tails that stretch across the sky (plate 14), comets are still widely represented by scientists as presenting no danger. Their masses are said to be so inconsequential in the Solar System scheme of things that a collision with a comet is 'like an insect trying to knock down a locomotive'; and in any case 'millions of years may pass before the occurrence of a fall of any importance. We may thus live on without worrying about this highly infrequent danger'.[75] Such statements do not accord with history. And as we shall soon see, they do not accord with science either.

The fallacy in the argument is that not all comets are small and insignificant. Most comets are only a few kilometres in diameter, the larger ones being progressively rarer. However, although very large comets are extremely infrequent, they are so massive that they dominate the entire comet system. If a random sample of a hundred comets were taken from the Sun's comet cloud, half their total mass would be contained in the largest one or two. In considering the effects of the fluctuating rate at which comets arrive in the planetary system, therefore, it has to be appreciated that the influx of mass is overwhelmingly dominated by the rarest monsters. This seems an obvious enough point, but it is continually overlooked, as a result of which a fundamentally erroneous view of the astronomical environment has been widely adopted by geologists, palaeontologists and others now attempting to interpret Earth history in the new catastrophist mould.

Comets more than about 50 kilometres in diameter will come in

Plate 14. The Great Comet of 1843, as drawn by Piazzi Smyth in March 1843, at the Cape of Good Hope (the comet was not visible from Europe). It appears that such comets, moving along the zodiac in short-period orbits, were a regular part of the night sky in early historical times (reproduced by kind permission of the Don Africana Library, Durban, South Africa).

from beyond Saturn and be deflected into unstable short-period orbits at intervals of about fifty thousand years. Many of these will eventually be thrown out of the planetary system – to wander perhaps in isolation through the Galaxy forever – but many will also be caught up by Jupiter's powerful gravity and thrown into short-period orbits that regularly intersect the Earth's path around the Sun. Over the few million years of a comet shower therefore, or when the Sun is deeply immersed in a spiral arm, perhaps twenty or thirty giant comets will join the inner planets, there to be destroyed by the heat of the Sun. Every 100,000 years or so therefore, during a high-risk period, the Earth encounters the debris left in the inner Solar System by a disintegrating giant comet. Most of this debris is in practice generated for only a fraction of the time before it is eventually blown away by the solar wind. But it is in fact during these periods when much of the freshly formed debris still circulates that the Earth is now known to be particularly vulnerable. It is obviously necessary that we learn a little more about how giant comets behave.

An important facet of comet evolution is their tendency to split into fragments, each fragment more or less temporarily leading an independent life as a comet, with head and tail. The splitting of a comet ('a flaming torch of exceptional size' according to Diodorus Siculus) was allegedly reported in 372 BC by the Greek Ephorus, each 'star' following a separate route over the sky. It has been suggested that Ephorus may have invented the splitting to give credence to a theory that comets were formed from the union of stars, but it now seems much more likely that the 'union of stars' theory was derived from ancient observations of comets splitting. Certainly Democritus reported that comets sometimes disintegrated into stars. Similar observations are found, scattered through the ages, in the records of the Chinese and Western astronomers, comets sometimes dividing into four or five. Thus there is no question nowadays that comets have a not infrequent tendency to split.

One of the most interesting comets observed to break up since the keeping of reliable records was that named after an Austrian army officer Biela in 1826 (although Pons, a janitor at the Marseilles Observatory, had seen it in 1805). The comet had a period of about seven years with an orbit which came within 20,000 miles of the Earth's, the Earth being nearest to the Biela orbit on 27 November each year. During a subsequent return, on 13 January 1846, the comet split into two under the eyes of the observers, a faint companion detaching itself and rapidly growing in brightness. The two comets were seen again on their return in September 1852, being then 1,500,000 miles apart. But in 1865, when the comet pair were due back, nothing was found: comet and companion had vanished. And then on 27 November 1872, the Earth ran into a swarm of meteors: 'they fell in large flakes, the streaks of fire coming down vertically in shower after shower with blinding balls of light and noiseless explosions looking like cascades of fireworks'. It is

Plate 15. Appearance of the sky during the night of 27 November 1872 due to the meteor shower resulting from the fragmentation of Comet Biela. Note the conspicuous bolide (from 'Les Etoiles Filantes' by Amédée Guillemin; Libraire Hachette, Paris, 1889).

estimated that 160,000 shooting stars came in over the six hour duration of the shower (plate 15). The Andromedid shower, as it is called, was seen again on 27 November 1885, and can be seen annually, much reduced, to the present day, the meteors having spread around the entire cometary orbit. Biela's comet, however, has never been seen again.[76]

Splitting is readily understood in the case of the giant parent of the Kreutz group,[77] so-called, which skims the surface of the Sun: comet material is very weak and the tidal forces exerted by the Sun's gravitational field are enough to generate a cascade of fragmentations. However this externally forced splitting is only a minor factor. Most splittings have occurred more or less anywhere along the orbital track of a comet, and it is clear that some other energy source must be involved. The nature of the source is unknown; it might for example represent stored chemical energy or the impact of a boulder in space. Thermal stresses, too, are adequate to crack cometary ice, creating surface rubble and systems of cracks which may be up to a hundred metres deep. Any fragmentation then may be spontaneous, or it may be triggered by an explosive encounter with a very small piece of Solar System debris.

Does it follow that all comets just evaporate away, leaving nothing more than dust orbiting the Sun? It would seem not. There are now several examples of comets that have an almost asteroidal appearance in the telescope, as if they had degassed and been left with a pumice-like surface. A number of asteroids moreover are in comet-like orbits, so the impression has grown that a comet can evolve not just by releasing gas and dust and running through a hierarchy of fragmentations, but that it can also degenerate into one or more inactive solid bodies. This has been confirmed by the recent discovery that most Earth-crossing asteroids are shedding meteors along their orbital paths.

Halley's Comet entered the domain of the inner planets in 1986, and was intercepted at around 70 kilometres a second by a small armada of spacecraft. Most of the outgassing observed from this comet (plate 9) during the flyby of the Giotto and Vega space probes came from only a few jets on the exposed surface, the rest of the nucleus having a black, solid appearance. It is possible that the comet was once active over the whole of its surface, say 20,000 or more years ago, but that the escaping gas has been gradually choked off by large dust particles falling back on to the nucleus. In years to come, the comet will be to all intents and purposes an asteroid. But it may still have an icy core. Not all comets appear to be changing into asteroids however, and it is likely that smaller or less dusty ones just evaporate away.

Larger comets, like several planetary satellites of similar dimensions, could possibly have rocky interiors. Whether or not this is so, we may expect that a giant comet, once it is captured into a short-period orbit, will over several thousand years generate a vast meteor stream along its orbital track until at an advanced stage of its evolution, perhaps ten to twenty thousand years into its lifetime, it has turned into an asteroidal body of still substantial size. From then on, there is a less conspicuous but extended period of decline, perhaps two or three times as long, during which the fragile core of the giant comet is whittled away, in part by encounters with small pieces of debris drifting around the Solar System. The result of this destruction is to create a huge expanding swarm of asteroidal flotsam and jetsam concentrated around the core and intermittently replenished from it.

The passage of the Earth through one such swarm was confirmed with the aid of lunar seismometers in late June 1975. This concentration of material was encountered as the Earth passed through a meteor stream (figure 6). As many one-ton boulders struck the Moon over five days as had hit it over the previous five years. A duration of passage of five days corresponds to a swarm of diameter at least 15 million kilometres. As the meteor stream in question, the Beta Taurids, strikes the day-time hemisphere, nothing was seen from the

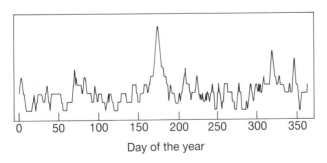

Day of the year

Figure 6. Diagram showing the average number of meteoroids striking the Moon on each day of the year over the period 1972–76, as recorded with seismometers left by the Apollo astronauts. The conspicuous peak at the end of June represents the flux over a few days in 1975 due to a swarm of more massive meteoroids coinciding with the Beta Taurid stream.

Earth, but a night-time encounter of the Earth with this material could have given rise to a spectacular fireball shower of several days' duration. If, however, the Earth had encountered this material at an early stage of the swarm's creation, when its diameter was say 1 per cent of its 1975 value, then although the encounter would have endured for only a few hours, it would have yielded a fireball impact rate *a million times higher*. This is the likely cause of the rare day-time fireball swarms recorded by the Chinese chroniclers. In addition, there would be a greatly enhanced probability of being struck by much larger bodies. In the aftermath of a major splitting or a major fragmentation, a giant comet or its asteroidal remnant is likely to be at its most dangerous. For a period of hundreds of years, a huge concentrated population of boulders and larger bodies will be present in the swarm with a greatly enhanced risk of planetary encounters. With the passage of time of course, the individual bodies will further disintegrate and the dust will spread away from its parent orbit so as to form a fat elliptical tube penetrated by the inner planetary system; later still, a few thousand years after the fragmentation, the dust will have spread into a flattish disc, the zodiacal cloud, occupying the whole domain of the terrestrial planets.

The astronomical framework for a theory of terrestrial catastrophism is now essentially complete, and it provides a perspective that is very different from the one that has dominated scientific thinking over the last three hundred years. It is also very different from the picture currently adopted by many modern catastrophists in the Earth sciences who, recently converted by hard geological evidence of astronomical inputs in particular strata, still ignore the hard astronomical evidence obtained by looking upwards.

This ignoring of basic data has two obvious outcomes. One is that discussion of catastrophism in geological circles is even yet largely conditioned by a view of Earth history which is essentially uniformitarian, with single, giant impacts occasionally interrupting a basically internal evolution. Such matters as the possibility of galactic cycles in the terrestrial record, or the climatic effects of stratospheric dusting from the debris of giant comets, are excluded by fiat. Another outcome is that catastrophe from the sky is assumed to be something that happens only on the vast timescales one meets in the geological record; the concept of astronomical catastrophe on historical timescales is effectively written off at the outset. We can illustrate these points by considering the Sun's orbit in the Galaxy and its encounters with massive interstellar nebulae. These latter, as we have seen, are distributed in a flat plane more or less coincident with the Milky Way. In relation to the disc of the Galaxy, and in common with the other stars in the disc, the Sun has two components of motion; one an up-and-down oscillation carrying it through the plane of the disc every 30 million years or so, the other an in-and-out motion with respect to the galactic centre. This latter 'galactic year' is about 250 million years long. Now stars and molecular clouds are concentrated towards both the plane of the disc, towards the inner regions of the Galaxy, and in spiral arms. The Solar System therefore encounters nebulae preferentially in these regions. The cumulative effect of these encounters is to impose long-term variations on the comet flux, with timescales of around 30 and 250 million years. The longer timescale is not strictly periodic since both the solar orbit and the presence of molecular clouds in the Galaxy may well be irregular; but the shorter 30 million year timescale, due to the galactic tide, should be more nearly a true period.

Consider now the situation when the Sun is inside a spiral arm. When the Sun is in the centre of the arm, and so also in the plane of the Galaxy, the tidal stress acting on the comet cloud is at a maximum, and the comet influx is at its greatest. The effect of the up-and-down motion in the Galaxy during any spiral arm passage is to impose a shallow porpoise-like motion on the Sun which causes it to coast along just above and just below the arm between successive penetrations. Whilst lingering at these levels the tidal stress again reaches a maximum, and so at the points of entry and departure from the spiral arm the galactic tide is reduced, causing the comet flux to decline. The effect therefore is to interrupt the basic 30-million-year cycle, giving rise to what is essentially a 15-million-year one. The effect may therefore be to force a 15-million periodicity on the Earth, with weak and strong cycles alternating. In a poorly determined terrestrial record, this might be misinterpreted as a 30-million-year one. Further, it is likely the comet cloud is replenished when the Sun passes through the spiral arms of the Galaxy. Comet activity on the Earth may be appreciable only during such passages, and if so, the terrestrial record will show a 15-million-year cycle, but one which is intermittent, switching on for 50 to 100 million years while the Sun is immersed in a spiral arm, and declining for a similar period outside it. If this galactic theory is correct, Earth history may therefore have

an episodic character, with 15- or 30-million-year cycles switching
on and off. The evidence suggests that this is exactly what is
observed (chapter 14). It should be said, however, that the galactic
environment is still not well explored. Several galactic periodicities
could be imposed on Earth processes, and their relative strengths
are still a matter for the empirical discovery of geologists and not the
revealed wisdom of astronomers.

A question of immediate and even human importance is where we
stand in relation to these periodic surges of comet activity. The Sun is
near the plane of the Galaxy and has also just emerged from the edge
of a spiral arm of the Galaxy, known as the Orion arm. Further, it is at
present moving through an environment which seems to be that of an
old disintegrating molecular cloud. There are numerous dense nebulae in the solar neighbourhood, most of which seem to be organized
into a ring of gas and young blue stars. The blue stars form a belt over
the night sky, noted by Ptolemy as long ago as AD 150, and known as
Gould's Belt. The ring is inclined at about 20 degrees to the plane of
the Galaxy; the Orion nebula to the south, and stars in the constellation Scorpio in the north, are part of this complex of material. This
massive ring of stars and gas is expanding rapidly, following some
energetic event of 30 million years ago, and only about 6 to 9 million
years ago the Sun passed through the rim, moving at some 20 to 25
kilometres per second. Because of the mass of material in this ring,
the comet cloud must have been disturbed during this close encounter. Once disturbed, the comets at the outer edge of the comet cloud
fall into the inner planetary system in about 3 to 5 million years. We
expect, therefore, that the Earth is in the declining tail of a comet
shower which peaked 3 to 5 million years ago. Thus the conditions
which would yield an exceptional flux of comets on to the Earth –
positioning near the galactic plane, proximity to a spiral arm, and
recent passage through a system of molecular clouds – are all simultaneously met by the Solar System at the present time.

As noted, very few comets present any danger to the Earth but the
occasional giant comet that is trapped in an Earth-crossing orbit of
short period rapidly develops in such a manner as to intermittently
generate a substantial swarm of bodies which are commonly
asteroidal. The giant comets recur at intervals of about 100,000 years
during a shower whilst the terrestrial encounters with the asteroidal
swarms, which may continue for a century or two, recur at intervals
of about a millennium.

To summarize: we expect both galactic periodicities and irregular cometary surges to be imprinted on the terrestrial record. Of the periodicities, one of around 30 million years (or 15 million years in particular circumstances) is likely to be the strongest. Because of the Sun's position near to both the Orion spiral arm and the plane of the Galaxy, we expect that the Earth is even now in a period of enhanced risk. The risk is further enhanced by the fact that the Sun has recently passed through a complex of debris associated with Gould's Belt.

We should look therefore for a currently disturbed comet cloud and a currently disturbed Earth. Such evidence exists (table 5). We expect the disturbances of the Earth and its biosphere to be profound: as we shall see, the extinction rate, the incidence of mountain building, the rise and fall of oceans, reversals of the Earth's magnetic field, are all expected to be under the control of the Galaxy. During peaks of activity, with a large disintegrating comet in the inner planetary system, we expect climatic recessions at random intervals of around a thousand years reaching peak intensity every hundred thousand years or so. This is an average picture however and more has to be known of recent and forthcoming giant comets to calculate the Earth's immediate past and future with any precision. Nevertheless, it follows that the Earth's environment must be *currently* hazardous because of the intermittent presence of great comets. There is strong evidence, indeed, that the last giant comet entered an Earth-crossing orbit only a few tens of thousands of years ago, so its asteroidal debris (including its resultant zodiacal cloud)

Table 5 Phenomena indicating recent disturbance of comet cloud

Phenomenon	Timescale
Persistence time of 'galactic alignments' in comet system	10 Myr
Characteristic 'relaxation time' of observed long period comet system	3—9 Myr
Arrival of comets into planetary system following passage through Gould's Belt	0—3 Myr
Onset of Pleistocene glaciation	2.5 Myr
Onset of worldwide vulcanism	2 Myr
Recent impacts (Australite and Ivory Coast tektites)	1 Myr
Taurid system (erstwhile giant comet)	0.01 Myr
Chiron (probably giant comet in an unstable orbit)	−0.01 Myr

are in orbit even now. The comet should therefore indeed have made its mark on history as well as on recent geology. Furthermore, a cometary body at least 250 kilometres in diameter, Chiron by name, has already been sighted in a chaotic orbit out beyond Saturn and there is good reason to believe it may be entering an Earth-crossing orbit within the next 100,000 years: Chiron may then be responsible for the next major ice age. Chiron is obviously not of great urgency but the same cannot be said of the asteroidal remnants of the most recent giant comet. Unfortunately most of these Earth-crossing bodies are completely uncharted and we have no means of knowing which of these will produce the next significant encounter with the Earth.

The Pleistocene ice epoch began about 2.5 million years ago, consistent with a sudden influx of comets from the disturbed cloud, although mountain glaciers were beginning to extend even before then. This epoch, like others which have occurred in Earth history, is broken up into a series of very erratic ice ages, each of duration typically 10,000 to 100,000 years. Again, these timescales are about those associated with the intermittent arrival and disintegration of the greatest comets, although no great precision can be put on the figures. The last ice age ended just over 10,000 years ago. We are therefore, at the moment, living in a warm interglacial period. Interspersing such warm periods, however, are short cooling episodes, very sudden in onset, and lasting typically ten to a hundred years. During these episodes the Earth's climate plunges into glacial severity. Such cooling episodes may be of great human consequence and will be discussed presently; for the moment we note only that for the first time there is a scientific basis for a catastrophist inter-pretation of human history; and we begin to discern that the threat from the sky may have a wintry aspect. Let us take a closer look at the material orbiting in the space around us.

10

Cosmic Swarms

BETWEEN 3 and 15 November each year, the Earth runs through a meteor stream coming from a small area in the constellation of Taurus the Bull. A young stream, made up of meteors not long released from their parent comet, would be narrow and probably strong, and the Earth would pass through it in half a day or so. The Geminid meteor stream is one such, the Earth encountering it on 10 December each year and the shower being over by 12 December. The Taurid meteor stream, on the other hand, takes the Earth 12 days to cross, and it is not very intense; it 'looks old'. It has two main branches, northern and southern, emerging from slightly different parts of the constellation, the branches being of about equal intensity.

The Earth runs into the same meteor complex in the summer but the meteors, striking the day-time hemisphere of the planet, are not then visible to the naked eye. This day-time stream, the so-called Beta Taurids, was discovered with radar using the Jodrell Bank radio telescope in the 1950s. The Beta Taurids come in between 24 June and 6 July with a sharp peak on 30 June each year. The orbits of these day-time Taurids match those of the southern Taurids closely. At first sight it seems that there is nothing exceptional about the Taurids; they are only one of seven or eight strong meteor streams which occur at various times throughout the year, and they are not particularly intense. To the casual observer the only noticeable feature of the stream is the high proportion of fireballs it produces, bright meteors which light up the landscape. Again, this is an indication of age, smaller meteors in the stream having long ago been destroyed in interplanetary space by processes of erosion.

Orbiting with the Taurid meteor stream is a small comet, discovered in 1818 by Pons. An orbit was calculated for this comet by Johann Encke, a pupil of the famous mathematician Gauss who had developed a method for calculating the orbits of bodies from their movement across the sky. Encke not only found an orbit, he showed that the comet was the same as one detected by Mechain in 1786, Caroline Herschel in 1795 and Pons himself in 1805. Encke predicted that the comet would return in May 1822 and in due course its return was observed, from Australia. There are interesting historical parallels with Halley's Comet; in both cases the comet is named, not after the discoverer, but after the man who calculated the orbit; Encke's Comet was the first instance after Halley of the recognized return of a comet; and, like Halley's Comet, it established the existence of a new class of object. Its orbit, in fact, is unique amongst the known comets.

There are about a hundred known short-period comets (say with orbital periods less than about 12 years as against a few million years for the long-period ones). All of these, with the sole exception of Encke's Comet, are in orbits which sooner or later lead to a close encounter with Jupiter and ejection from the Solar System by the powerful gravitational pull of that planet. But Encke's Comet has somehow established itself in a stable orbit such that it never comes within the Jovian sphere of influence. It does however come close to the Earth. In addition there may be one or two thousand asteroids over a kilometre across in orbits which make them a potential collision hazard with the Earth. None is more than about 10 kilometres across, their surfaces are extremely dark, and fewer than fifty have been discovered. They are known as Apollo asteroids after the prototype discovered in 1932. Encke's Comet is in fact the sole example of an active comet in an Apollo orbit. Without decay or collision, the comet would probably continue in orbit for tens of millions of years. The orbit is highly elongated (figure 7) and the comet may approach to within 0.34 astronomical units of the Sun, when its surface temperature reaches 450 degrees centigrade, and go out to 4.1 astronomical units, when its temperature is −120 degrees centigrade. The comet's orbit is inclined at only 12 degrees to that of the Earth's. This low inclination and the shortness of the orbital period, ensure that there is frequent interaction between the Earth and the material of the Taurid stream, with the potential for occasional very close encounters with Encke's Comet. On the face of it, then, the Taurids seem to be an elderly, diffuse meteor stream with no peculiar characteristics other than their association with a comet, in an unusual orbit, and just too faint to be seen with the naked eye. There is much more to the Taurid stream, however, than this.

In a study of fireball records from the first to the fifteenth centuries AD, the astronomers Astapovic and Terenteva[78] found that the fireball flux from the Taurids was much more intense a thousand years ago (see figure 4(b)), indeed the northern stream was then the stronger of the two. The Soviet astronomers found that 'Taurids were the most powerful shower of the year in the eleventh century (with 42 fireballs belonging to them) and no shower, not even the great ones, could be compared with them as to activity'. In the not too distant past, therefore, the Taurids were revealing themselves as outstanding performers in the night sky.

Two recent discoveries confirm the suspicion that the stream is a remarkable one. The first of these is that, amongst the 80 or so known Apollos, 6 or 7 are orbiting within the Taurid meteor stream (table 6). There are some interesting objects in table 6. Hephaistos, for example, is the largest known Apollo asteroid, being about 10 kilometres in diameter. Its orbit closely resembles that of the comet's in every respect but one: the long axis is roughly at right angles to that of the Encke orbit. The orientations of these orbits change very slowly, only a few degrees in a millennium. It could be that Encke's Comet and Hephaistos were once one body, but if so they must have separated over 20,000 years ago. Then there is Oljato, a dark asteroid about 1.5 kilometres across. This body shows visual evidence, through the telescope, of slight outgassing. Further, when the Pioneer Venus probe passed near the asteroid on several occasions in 1982 and 1983, magnetometers on board recorded sudden changes in the magnetic field, as if the probe were passing through magnetized gas in the wake of the asteroid.[79] It looks as if the 'asteroid' is in fact an almost defunct comet. Each of these bodies pursues an elliptical orbit around the Sun, but because of the influence of the planets, Jupiter and Saturn in particular, these orbits slowly evolve. The size and shape do not change greatly but the orientation of the orbit in space does. First, relative to the mean plane of the Solar System, the inclination of the orbit slowly oscillates up and down; but more significant is the orientation, in the Earth's orbital plane, of the orbit's long axis. If the asteroids were unconnected with each other and with Encke's Comet their long axes would be orientated randomly; but they are on the whole closely bunched. It can be shown that there is less than one chance in a million that even three asteroids would lie so close to the orbital track of Encke's Comet by chance. Now it is very unlikely that these bodies are the only ones

Table 6 Probable debris from the giant progenitor comet

Object	a (AU)	e	i (deg)	$\tilde{\omega}$ (deg)	Size of object
Meteor streams					
S Taurids	1.93	0.896	5.2	153.2	
N Taurids	2.59	0.861	2.4	162.3	
β Taurids	2.2	0.85	6	162.4	
ξ Perseids	1.6	0.79	0	137	
S Piscids	2.33	0.82	2	104	millimetres
N Piscids	2.06	0.80	3	130	
Sχ Orionids	2.18	0.78	7	180	
Nχ Orionids	2.22	0.79	2	179	
Active comets					
Encke	2.2	0.85	11.9	160	
Rudnicki	—	1.00	9.1	154.7	
Asteroids					
2201 Oljato	2.2	0.71	2.5	172	
1982 TA	2.2	0.76	11.8	128	
1984 KB	2.2	0.76	4.6	146	few kilometres
5025 P-L	4.20	0.895	6.2	145.8	
2212 Hephaistos	2.1	0.83	11.9	258	
1987 SB	2.16	0.650	2.9	167.4	
Unseen companion	2.4	0.86	—	160	
Impactors on the earth or moon					
Boulder flux (1971—7)					metres
Boulder swarm (1975)					metres
Tunguska object (1908)					100 metres
Bruno object (1178)					few kilometres
Larger complexes (dust)					
Stohl stream					millimetres
Zodiacal cloud					sub-millimetre
β Taurid 'trail'					millimetres or more

The orbital semi-major axis a is in astronomical units (1 au = Earth/Sun distance = 150 million km). e represents the eccentricity of the orbit (a circular orbit has $e = 0$, a highly elliptical one has e close to unity). i represents the orbital inclination in degrees, and $\tilde{\omega}$ measures, approximately, the orientation of the major axis of the orbit in the Earth's orbital plane. The general similarity of $\tilde{\omega}$ for these objects indicates that major fragmentation of the progenitor comet has since the main devolatilization ended about 10,000 years ago. There is evidence of vestigial cometary activity from Oljato. The inclusion of Comet Rudnicki and Asteroid 5025 P-L in this table has been justified recently by Olsson-Steel (1987).

orbiting within the Taurid meteors: discovery of the Apollo asteroids as a whole is only about 5 per cent complete. We are therefore led to the remarkable conclusion that there are between one and two hundred asteroids of more than a kilometre diameter orbiting within the Taurid meteor stream. It seems clear that *we are looking at debris from the breakup of an extremely large object*. The disintegration, or sequence of disintegrations, must have taken place within the past twenty or thirty thousand years as otherwise the asteroids would have spread around the inner planetary system and be no longer recognizable as a stream.

The second discovery, due mainly to the Czechoslovakian astronomer Stohl,[80] is that enveloping the Taurids, Comet Encke and these particular asteroids is a broad tube of meteoric debris. The Earth enters this tube in April and does not emerge again until about the end of June, entering again in October and re-emerging in December. It seems that about half the 'sporadic' meteors one sees – random shooting stars which do not seem to belong to any specific meteor stream – are not sporadic at all, but are part of this immense swathe of material, of which the Taurids and the bodies within it are the core. The Stohl stream is apparently double (figure 7) due probably to an exceptional fragmentation which, if the orbital backtracking of several Taurid meteors by Whipple and Hamid[81] is accepted, occurred around 2700 BC when the primary body encountered a stray body in the asteroid belt between Mars and Jupiter. This pronounced splitting into two streams could be significant in view of the dual character of divinity generally discussed in earlier chapters.

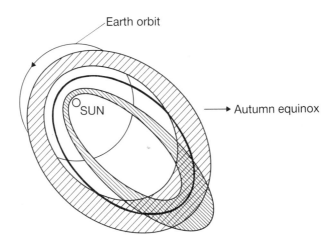

Figure 7. Diagram showing the orbit of Comet Encke (thick line corresponding also to the orbit of the Beta Taurids) and the two broad meteoroidal swarms discovered by the Czechoslovakian astronomer Stohl. The similarities between these orbits and the general merging of these streams with the meteoroidal complex and zodiacal cloud in the plane of the ecliptic suggest a common origin in the past from a single giant progenitor comet.

The mass of the meteoric material within the Stohl stream is 10 or 20 trillion grammes (we use trillion in the American sense of a million million times a million). The mass of the co-orbiting asteroids is likely to be about the same. Adding in the mass of gas and very fine dust which must have been blown away subsequently, we find that the original body must have been about 100 kilometres across. We have found our giant comet!

The existence, in the not too distant past, of a giant comet in an Apollo orbit also explains the existence of the zodiacal light. In the early evening, or just before sunrise, this delicate, beautiful pyramid of light can be seen rising out of the horizon, lying along the line of the zodiacal constellations. It is best seen in the tropics, where it rivals the Milky Way, but unlike the Milky Way it has a more regular, smooth appearance. We have seen that the zodiacal light is made up of light scattered from myriads of bodies orbiting between the inner planets. It is brightest close in to the Sun, and it could be that ancient Egyptian portrayals of the Sun as a winged disc refer to the zodiacal 'wings' extending from it. From the rate at which the Earth sweeps up meteors one finds that the mass of small meteoric material in the zodiacal cloud, say particles less than a millimetre diameter, is perhaps 20 to 40 billion tons. About half of this belongs to the Stohl stream. Now the particles orbiting within the cloud collide at a few kilometres per second, and break up, the microscopic debris eventually spiralling into the Sun. The zodiacal cloud loses between 10 and 40 tons of mass every second in this way, and without replenishment it would vanish in about fifty or a hundred thousand years. And yet, there are no adequate sources of replenishment for this dust at the present time.[82] Encke's Comet is the major known supplier of dust in the inner planetary system at present, but even so, it loses only an average of a ton a second of material over an orbit, only 0.3 tons per second of which is dust; it cannot even remotely resupply either Stohl stream or zodiacal cloud. Once more, there seems to be no avoiding the conclusion that the progenitor of Comet Encke and the Stohl stream must have been a truly exceptional body, in the same class as the great historical comets. And it must have been pouring out dust copiously in the not too distant past, since the Stohl stream, diffusing into the zodiacal cloud as a whole, would not stay recognizable for more than twenty thousand years or so.

Backtracking the orbits of Encke and Oljato, we find that 9,500 years ago the orbits were nearly identical. It is possible there was a major disintegration of the prime body then, with much debris created of which Comet Encke and Oljato are the largest known bodies, followed by similar disintegrations of the other comets and asteroids of the stream. Oljato itself is in an orbit which brought it virtually into the Earth's orbital plane for some centuries around 3000–3500 BC, ensuring a series of very close encounters with the Earth over that period. If, as is very likely, it was then an active comet, it must have been a spectacular zodiacal traveller, whose cycles of coming and going, flaring and fading, must have made it a ·perplexing object in its own right, even if overshadowed by the more formidable Encke. The meteor orbits, traced back in time, seem also to come together at particular epochs. The present-day northern Taurids, for example, may have broken away from Encke's Comet or a Taurid asteroid about a thousand years ago, consistently with the surge of activity recorded in the sky around that time, particularly by the Chinese astronomers (figure 4). The whole complex therefore seems to be undergoing an avalanching self-destruction as the debris accumulate and collide. Orbital backtracking of the Taurid meteors and asteroids thus reveals a system which has been evolving actively within the timescale of civilization.

This unique complex of debris is undoubtedly the greatest collision hazard facing the Earth at the present time. It is likely that hundreds of thousands of bodies, each capable of yielding a multi-megaton explosion on Earth, are orbiting within the stream. Such bodies are expected from the progressive disintegration of the asteroids – inactive comets – within it. From time to time collisions will remove some proportion of the dusty crust from an asteroid, exposing its icy interior. It will then become a comet for a few centuries. Probably, also, there are one or more concentrations of material – temporary swarms – arising from the rapid disintegration of one or two dominant but yet undiscovered bodies within the Taurid stream. The Moon, as we have noted already, ran through such a swarm between 22–26 June 1975. Lunar seismometers left by the Apollo astronauts recorded the impacts of a swarm of ton-sized boulders, as many hitting the Moon over those five days as had struck it over the previous five years (figure 6).

We seem to have found, then, the vital element missing from the works of the early Biblical catastrophists – Whiston, Radlof, Donnelly, Velikovsky and the rest – namely, a scientific rationale, a relatively secure astronomical framework. Biblical and geological catastrophism are, after all, inextricably linked. While this clearly justifies an urgent reappraisal of the ancient tales of celestial catastrophe, the new information is extremely awkward for a generation of astronomers who insisted that Velikovsky was no more than an erudite charlatan. Astronomers, indeed scientists generally, like to think of themselves as tolerant judges and very adaptable to fresh discoveries. The evidence in this instance is however mostly the other way. One may therefore expect that in some circles the data now emerging from the Taurid meteor stream will be ignored in the hope that something reassuring will turn up. While this is a time-honoured scholarly ploy for the handling of discordant new facts, there is a moral dimension in this instance: the swarm has teeth. As to how sharp these teeth can be, we shall now see by looking at two impacts which have taken place during the present millennium.

I I

Close Encounters

JUST after 7.15 am local time on 30 June 1908, in the central Siberian plateau, there took place an impact of ferocious intensity. Yet so isolated and vast is this region (half as large again as the USA), it was almost twenty years before the Western world became aware of the event.[83]

On the night of 30 June and 1 July, the sky throughout Europe was strangely bright. Throughout the United Kingdom, over 3000 miles from the point of impact, it was possible to play cricket and read newspapers by the glow from the night sky. From Belgium came descriptions of a huge red glow over the horizon, after sunset, as if a great fire was raging. This strangely bright sky was seen throughout Europe, European Russia, Western Siberia and as far south as the Caucasus mountains. Photographs were taken at midnight or later, with exposures of about a minute, in Sweden, in Scotland, and as far east as the university city of Kazan, on the banks of the river Volga. And yet the Urals, dividing Europe from Asia, are 400 miles east of Kazan, and the impact point was over 1200 miles beyond that again.

Much comment was excited in newspapers and learned journals at the time. Some thought that icy particles had somehow formed high in the atmosphere and were reflecting sunlight. Others considered that a strange auroral disturbance was involved. The Danish astronomer Kohl drew attention to the fact that several very large meteors had recently been observed over Denmark and thought that comet dust in the high atmosphere might account for the phenomenon. But there was no agreement as to what had happened.

Over 500 miles to the south of the fall, a seismograph in the city of Irkutsk near Lake Baikal, close to the Mongolian border, registered strong earth tremors.

Nearly 400 miles south-west of the explosion, at 7.17 am on 30 June, a train driver on the Trans-Siberian express had to halt the train for fear of derailment due to the tremors and commotion.

Fierce gusts of wind were felt in towns 300 to 400 miles away.

In an Irkutsk newspaper dated 2 July it was reported that, in a village more than 200 miles from the Tunguska river, peasants had seen a fireball brighter than the sun approach the ground, followed by a huge cloud of black smoke, a forked tongue of flame and a loud crash as if from gunfire. 'All the villagers ran into the street in panic. The old women wept and everyone thought the end of the world was approaching.'

Almost twenty years later, a farmer recalled that, 125 miles from the point of impact:

When I sat down to have my breakfast beside my plough, I heard sudden bangs, as if from gunfire. My horse fell on its knees. From the north side above the forest a flame shot up. Then I saw that the fir forest had been bent over by the wind and I thought of a hurricane. I seized hold of my plough with both hands so that it would not be carried away. The wind was so strong that it carried off some of the soil from the surface of the ground, and then the hurricane drove a wall of water up the Angora.

From another village at the same distance out, the local newspaper reported that:

A noise as from a strong wind was heard, followed by a fearful crash, accompanied by a subterranean shock. . . . This was followed by two equally forceful blows and an extraordinary underground roar like the sound of a number of trains passing simultaneously over rails . . . a heavenly body of fiery appearance cut across the sky . . . when the flying object touched the horizon, a huge flame shot up that cut the sky in two.

Forty miles out, at the Vanavara trading post, a farmer was sitting on the porch of his home. He reported:

I saw a huge fireball that covered an enormous part of the sky . . . the whole northern sky appeared to be covered with fire. I felt a great heat as if my shirt had caught fire. Afterwards it became dark and at the same time I felt an explosion that threw me several feet from the porch. I lost consciousness.

At this distance too, 'the ground shook and an incredibly long pronounced roaring was heard. Everything round about was shrouded in smoke and fog from the burning and fallen trees'.

The closest eyewitness description came from a family in a tent 25 miles south-east of the epicentre. The family had been thrown in the air and several were knocked out. When they recovered they found the forest ablaze around them and much of it devastated. It is not unlikely that some nomadic herdsmen were closer to the centre of the impact; if so, they would have perished in the blast. A thousand reindeer may have been killed as well.

Local Siberian newspapers carried stories of a fireball in the sky, and a fearful explosion, but by the autumn of 1908 these stories had died out, and they went unnoticed in St Petersburg (Leningrad), Moscow and the west. The region was arguably one of the most inaccessible places on Earth, in the centre of Siberia. It comprises swamps, huge pine forests and peat bogs. It was populated only by nomadic reindeer herdsmen (the Tungus, now renamed Evenki) and large, ferocious mosquitoes. However rumours of an extraordinary event persisted, transmitted back by geologists and other intrepid researchers working in the area. These attracted the attention of a meteorite researcher, Leonard Kulik, who in 1921 led an expedition travelling from Petrograd (Leningrad) on the Trans-Siberian express across the Urals into Siberia, stopping finally at Kansk. Here he learned that the impact must have taken place hundreds of miles to the north, near the Tunguska river. It was not until 1927 that an expedition, again led by Kulik, finally penetrated to the site of the 1908 explosion.

Through this expedition, and others which followed, the event may be recreated. The fireball was blindingly bright, even dimming the Sun in comparison, and was seen over an area almost 1,000 miles across. It approached on a shallow trajectory from a roughly southern direction, leaving a thick dust trail and crossing the sky in a few seconds. After it had disappeared from the sky, a vertical column of fire was seen over the impact site, and was visible from as far away as 300 miles. Deafening explosions were heard, followed by thunder and rumbling, the explosion being heard over a circular area almost 900 miles across, the concussion wave being enough to throw many witnesses to the ground. Over this area also buildings were shaken by ground tremors. (It was later learned that seismographs had detected these as far away as the USA and Java.) The object seems to have disintegrated at about 5 miles altitude, halted by the atmosphere, effectively converting its energy of motion into heat and so creating a ball of fire. It broke up over an area known as the Southern Swamp, a depression 4 to 7 miles across consisting of peat bogs and swamps. Surrounding this area was forest, flattened and uprooted by the blast to a distance of about 40 miles, although the shock waves extended almost 70 miles out. Some trees remained standing, stripped of their branches. Out to a distance between 10 and 20 miles from the epicentre, brushwood and trees had been scorched on their inward facing sides.

The energy of the explosion has been calculated from the extent of the flattened forest, and from the small pressure waves which arrived at the speed of sound and were recorded on barographs around the world,[84] the latter calculation being the more reliable. The wave trains were unlike any others which had been recorded up until that time but resemble those obtained from a hydrogen bomb explosion (figure 8). It seems that the impact had an energy of 30 to 40 megatons, about that from a few dozen ordinary hydrogen bombs, and similar to the Boulder City impact of the prologue.

The date of fall (30 June) corresponds to the passage of the Earth through the maximum of the Beta Taurid stream.[85] From this and its trajectory, it appears that the Tunguska object was part of the Taurid complex. Probably, the Earth passed through a swarm within the stream. The occurrence, this century, of an impact with the energy of a hydrogen bomb does give cause for some concern, and it is

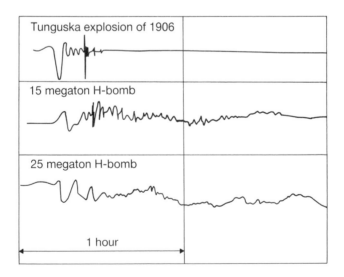

Figure 8. Pressure waves from the Tunguska explosion detected in south-east England 5720 km from the impact. For comparison, pressure waves from hydrogen bomb explosions are shown also: Bikini to barograph at Aspendale 5920 km, and Novaya Zemlaya to Aspendale 14000 km. From this an impact energy of about 30 megatons can be inferred. From E. L. Deacon (note 84).

interesting to speculate on whether one's historical perceptions would be quite the same had the bolide struck an urban area or a city. As it happens, however, the Tunguska impact is fairly trivial:

In this year, on the Sunday before the Feast of St. John the Baptist, after sunset when the moon had first become visible a marvellous phenomenon was witnessed by some five or more men who were sitting there facing the moon. Now there was a bright new moon, and as usual in that phase its horns were tilted toward the east; and suddenly the upper horn split in two. From the midpoint of the division a flaming torch sprang up, spewing out, over a considerable distance, fire, hot coals, and sparks. Meanwhile the body of the moon which was below writhed, as it were, in anxiety, and, to put it in the words of those who reported it to me and saw it with their own eyes, the moon throbbed like a wounded snake. Afterwards it resumed its proper state. This phenomenon was repeated a dozen times or more, the flame assuming various twisting shapes at random and then returning to normal. Then after these transformations the moon from horn to horn, that is along its whole length, took on a blackish appearance. The present writer was given this report by men who saw it with their own eyes, and are prepared to stake their honour on an oath that they have made no addition or falsification in the above narrative.

This curious report is written in the chronicles of the medieval monk known as Gervase of Canterbury. The year of the event was AD 1178 and the date, 18 June on the so-called Julian calendar, converts to the evening of 25 June on the modern Gregorian one. If real, it is clear that some extraordinary event on the Moon is being described, and the meteorite expert Hartung[86] proposed that what was observed and recorded 800 years ago was the impact of a body on the Moon. The flame, he suggested, was the writhing of incandescent gases, or sunlight reflecting from dust thrown out of the crater. The blackish appearance of the Moon along its whole length was a temporary suspension of dust buoyed up by a transient atmosphere.

Of course, a tale recorded by a medieval monk, by itself, need not signify much, although the date is suggestive. But it happens that in this case there is a strong physical evidence in support. The Moon's surface is sprinkled with impact craters, fossil remnants of past collisions with small bodies; and because the lunar surface is very old, the craters represent the accumulated history of four billion years of bombardment. Hartung deduced that if there was a crater, it would be at least 7 miles in diameter, possess bright rays extending from it for at least seventy miles, and would lie between 30° and 60°

north, 75° and 105° east on the Moon. These co-ordinates are the equivalent of latitude and longitude on Earth and are measured, roughly, from the nearest point of the Moon facing the Earth. The crater would therefore lie very near the edge of the Moon or possibly be just over it, on the far side.

As it happens, one crater with the predicted characteristics exists. There is a crater named after the seventeenth-century heretic Giordano Bruno (see plate 16). This crater is located 36° N and 105° E, within the predicted area. It is 13 miles in diameter and is distinguished by its remarkable brightness, and by the brilliant system of rays which extend several hundred miles out from it. Such

Plate 16. The Giordano Bruno crater on the Moon, photographed during the ill-fated Apollo 13 mission. Note the strikingly vivid (i.e. fresh) rays indicative of recent formation (by kind permission of the Johnson Space Center Mapping Science Laboratory).

rays are not uncommon around lunar craters; they are formed by debris thrown far beyond the crater rim during the explosive excavation. Any binoculars turned to the full Moon will show such a ray system radiating out from the young crater Tycho: this crater is about 40 miles across and its ray system extends over much of the facing hemisphere of the Moon. Giordano Bruno lies just beyond the edge of the Moon and its rays are not visible from the Earth. Nevertheless, Lunik and Apollo missions have revealed that the crater and its ray system are impressively brilliant, even although they come from a crater only a fraction of the size of Tycho. The significance is that bombardment by microscopic particles and gas atoms has not had time to dull the material around the crater. It has been very freshly exposed.

Further evidence in support of the identification was forthcoming from an investigation carried out by the lunar astronomers Callame and Mulholland.[87] They found that although the crater was 15° into the far side of the Moon the ejecta would be hurled such distances that 'the event would have been not only visible but sufficiently apocalyptic to have justified the description given in the Canterbury Chronicle'. They also found direct evidence in the form of a small natural vibration of the Moon. Although the Moon rotates so that the same face is always pointing towards the Earth, there are also small oscillations about this position. Some of these can be stimulated by an impact, in the same way that a weight suspended on a rope can be set swinging by a blow. A blow on the scale needed to excavate Giordano Bruno would have yielded a tiny wobble of the Moon about its polar axis, a surface movement of no more than a few metres, with a period of three years. Detection of such a movement by eye from 380,000 kilometres would be impossible even with the most powerful telescopes. There is, however, another way.

Since the early 1970s, laser ranging of the Moon has become possible because of reflectors left on the lunar surface by the Apollo and Lunik landing missions. Once a week since 1973, a laser beam is fired from the McDonald Observatory in Texas towards one of these reflectors and the arrival of the returning light is timed to about a billionth of a second. Several thousand such observations have been made over the years and much fine structure in the motion of the Moon has been found. Many new results have come from these observations, one of them being the discovery of a 15-metre oscillation of the lunar surface about its polar axis, with a period of about three years. This mode of vibration dies out over twenty thousand years or so, and the result can only be explained by a recent large impact, whose magnitude was about that required to form the Bruno crater.

Many studies of impact cratering have been carried out, and for a given size of crater it is possible to infer the energy of the impact which created it. It turns out that the Giordano Bruno crater was probably caused by a missile a mile or so across striking the Moon with an energy of about 100,000 megatons, or about ten times the combined nuclear arsenals of east and west.

The Giordano Bruno event took place in the century when the Taurid fireball flux was at a peak (see figure 4). It is remarkable that the terrestrial implications of this extraordinary event seem to have gone generally unrecognized.[88] The damage from blast and heat alone caused by such an impact on land would be far greater than that described in our opening story; probably, it would cause prompt destruction of life over continental dimensions; and probably too, it would have even more disastrous effects through a global climatic catastrophe. And yet the effective target area of the Moon is only about a twentieth that of the Earth, and the impact, if the interpretation given above is correct, took place less than a millennium ago, when the Taurid meteors were highly active. The basic thesis of this book is that the terrestrial environment is a good deal more hazardous than has been realized so far. In the Bruno impact, it would be hard to find a more dramatic confirmation.

At the moment of the Bruno impact the Moon was low in the sky over much of Europe, being visible east of a line stretching from Oslo, Stockholm, southern England, north-west France and Spain, and along the south-east coast of Brazil. It seems very likely that others saw and recorded the Bruno or similar events. It may be that, tucked away in the archives and libraries of the great European monasteries, further evidence of crucial importance still exists.

To sum up, we now know that over 22–26 June 1975, the Moon suffered an intense bombardment of boulders; that on 30 June 1908 a 30-megaton or so impact took place in northern Siberia; and that on 25 June 1178 an impact of energy about 100,000 megatons appears to have taken place on the Moon, during an epoch when the Taurid fireballs were highly active on Earth. And we now realize that these disasters from the sky are associated with swarms of asteroidal

12

Ancient Echoes

ARCHAEOLOGY, archaeoastronomy, mythology, classical history and so on have long been studied under the assumption that the night sky of, say, three thousand years ago was no different from the one we see now.[89] Certainly the slow movement of the celestial pole amongst the stars is generally recognized; also, past eclipses can be dated by astronomers to coincide with those known from ancient history sources. This gives a strong *prima facie* colour to the comfortable view of an orderly, stable universe. Let us summarize the astronomical evidence which reveals that this assumption can no longer be justified.

As little as a thousand years ago, we have seen, the sky was dominated by fireball activity emerging from the Taurus constellation, accompanied probably by an impact which, had it been terrestrial rather than lunar, would have made it unnecessary to write this book. Three or four thousand years ago, the whole Taurid complex must have been larger and more active. Orbiting within it were active comets, celestial objects travelling along the zodiac and easily visible to any shepherd tending his flock by night. Encke's Comet is dying rapidly and will become an asteroid within about 50 years; but in the century after its discovery the comet is known to have been seen as a naked eye object about a dozen times. Even today, if it were just a little brighter we would now have a ghostly zodiacal traveller coming and going in the night sky in a series of complicated cycles, as easily seen as the planets but a good deal more perplexing.

Going into the more remote past, we enter periods of history when the celestial fireworks from the cometary material were much more dramatic. Even in fairly recent historical times the Taurid stream has been active: this is evidenced, not only by the medieval observations but also by orbital backtracking of meteors and by the fact that there seems to be an intense meteoric core to the Taurid stream even material which emerge from time to time out of the constellation Taurus.

It is now abundantly clear that we are faced with a sky far removed from the harmless one with which we have been presented by generations of scholars. We are approaching the state where a quantitative assessment of the hazards facing us can be attempted. But first, armed with these new astronomical insights, let us take a fresh look at old evidence: let us test our new model against the ancient tales of celestial catastrophe.

today, manifesting itself as a swarm (figure 6). If the meteors were more than say 10,000 years old the stream would be unrecognizably dispersed; easily visible cometary activity has therefore, in all probability, been taking place throughout that period and beyond. The disintegration history of the giant comet itself probably covers tens of thousands of years. Alone or as part of a small number of major comets, it probably released enough dust to initiate the last ice age, a wintry drama *par excellence* which reached a peak around 20,000 years ago. Dusty streamers from the comet, lying in the plane of the zodiac, must have been easily associated with the main comet and its meteors.

In neolithic and early historical times, then, a string of naked-eye comets was moving along the zodiac much like the planets, still active debris from a hierarchy of disintegrations of the major body. Encke's comet could be the remnant core of the original body; or it could be merely one of the several hundred remnant Apollo asteroids which was re-activated for a few centuries, perhaps by a recent collision breaking through its dusty surface. At any one time, it is likely that there were one or a very few dominant bodies. It is expected that there were major splittings from these from time to time over the centuries. Some of the smaller fragments would be transient in the extreme; others, respectable comets in their own right, would persist for centuries or millennia. The slow evolution of the cometary orbits ensures that, at a few brief epochs in the past, the orbital tracks of the major comet and Earth intersected (see, for example, figure 9). Close encounters with a major, active fragment, over those centuries, must have been spectacular if not terrifying, with the comet nucleus, brighter than Venus, crossing the sky in a few hours, accompanied by a complex, striated red tail bisecting the sky. Meteor storms of ferocious intensity, recurring annually when the Earth crossed the debris of the comet, must have taken place during those epochs, the shooting stars blazing out from a small region of sky in Taurus or Aries. On some such occasions the sky must have been filled with brilliant fireballs, such storms lasting for several hours. Some of these fireballs would on occasion reach the upper atmosphere or the ground, and there would be explosions, there could be no doubting the association of these spectacular phenomena with the supreme being in the sky.

Such close encounters with the Earth, although spectacular, would however be rare. A comet increases rapidly in brightness as it

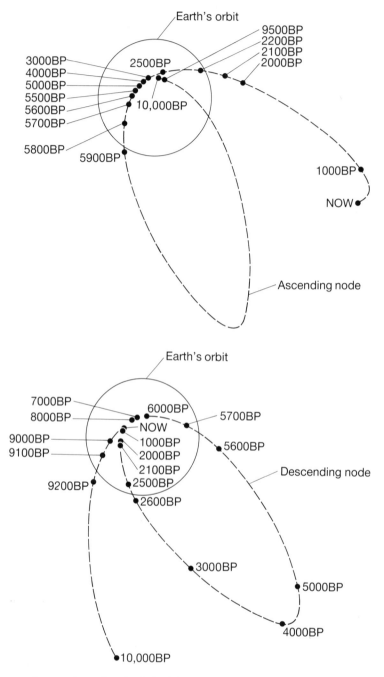

Figure 9. Intersection points (nodes) of Comet Encke's orbit with the plane containing the Earth's orbit, at various dates before the present (BP). During the first millenium BC up to the time of Christ, the orbit of the comet or that of its associated dust trail would have intersected that of the Earth, leading at this time to annual fireball storms of amazing intensity.

approaches the Sun, and a more usual manifestation of the cosmic debris would be the regular appearance of one of the major fragments as a spectacular object in the morning and evening sky.

As the millennia passed, the comets lost their volatile gases, their tails diminished and the comets eventually faded from sight. It is likely, however, that one or a few strong meteor showers remained, repeating annually and predictably. Of these, the Taurid meteor shower – really a complex of closely similar showers – was undoubtedly outstanding. Throughout the year an intense zodiacal light must have been visible, in the pre-dawn and post-dusk skies, as a pillar stretching upwards from the Sun. During the night, twin pillars would still be visible rising on each side from opposing horizons. Patches of light, due to concentrations of debris released in past disintegrations of the prime body, would seem to drift through the constellations of the zodiac. As we entered the pre-Christian era, even these signs gradually faded from sight until today, only a faint zodiacal light and an ancient, diffuse meteor stream remain as visible signs.

To summarize, we expect:

1. intense annual fireball activity emerging annually from the Taurus/Aries region, this activity reaching prodigious levels on occasion;
2. comets, with perhaps one or two dominating, regularly tracking along the zodiacal path in the night sky, with occasional spectacular enounters;
3. one or two comets regularly visible as a brilliant object in the morning and evening sky; and
4. pillars, patches and streams of light in the night sky, lying along a bright zodiac.

In table 7 we list some of the overlapping timescales of human and astronomical interest. It makes little sense (or so it seems to us) to try to understand the ancient traditions of celestial catastrophe while ignoring these new results. Such matters as the Bronze Age destructions and the traditions of crisis – the world coming to an end through celestial fire and so on – have to be looked at anew. We have no illusions about the problems involved in this programme: there is a heavy intellectual investment in prevailing interpretations and small adjustments to them will not necessarily do. Nevertheless, it seems that a cosmic message has been handed down through the

Table 7 Timescales relating to human development and the terrestrial environment

Phenomenon	Timescale*
Astronomical	
Giant comets injected into short-period orbits	100,000 yr
Orbital evolution of Chiron	100,000 yr
Significant evolution of zodiacal cloud	1,000—50,000 yr
Lifetime of giant comet	10,000—30,000+ yr
Interval since maximum extent of last ice age	18,000 yr
End of continental glaciation	10,000 yr
Lifetime of typical meteor stream	500—5,000 yr
Significant evolution of giant comet debris	1,000—5,000 yr
Significant evolution of Taurid meteor stream	1,000—5,000 yr
Lifetime of normal active comet	300—1,500 yr
Cultural	
Beginning of homo sapiens	40,000 yr
Beginning of food production and domestication	10,000 yr
First civilizations; irrigation; cuneiform texts	5,000 yr
Constellation building (main epoch)	5,500 yr
Oldest celestial myths	3,000—5,000 yr
Main epoch of megalith and ziggurat building	3,000—5,000 yr
Earliest calendars	3,000 yr
Well documented history	less than 2,000 yr

* Very approximate. The slowness of human development during the ice age contrasts noticeably with the speed of cultural evolution following its end. There is much overlap of astronomical and cultural timescales, and indeed the zodiacal cloud and Taurid meteors are visible even now, while Encke's comet is just below naked-eye visibility. In antiquity, with the complex of debris much more active, it is likely that the prime comet, lesser comets, and debris were regularly visible in the zodiac, and associated with fierce annual fireball showers and occasional impacts.

millennia; and the experts have failed to interpret the archaic code.

In seeking evidence of the ancient turbulent sky, we should not necessarily expect descriptions, perhaps four or five thousand years old, to be given in modern scientific terms. We do not seek comets tracking along the zodiac but gods wandering along a celestial river; rather than evanescent patches of light we should seek temporary celestial islands; rather than fireball swarms and a Tunguska impact we expect celestial thunderbolts hurled in anger; and in tales of the world's end, we should be alert for indications that celestial catastrophe came specifically out of the constellation Taurus the Bull, or in more remote epochs, because of the evolution of the meteor orbits, out of the neighbouring constellation Aries the Ram.

In interpreting ancient and frequently obscure texts dealing with the sky, and the evolution through the ages of ideas about the universe and its celestial gods, a proper methodological approach is required: it is incorrect to ignore the actual astronomical environment. Many interpretations of the gods and their adventures could be made to seem plausible, and in these tales have been seen visitors from outer space and planets hurtling past the Earth. This relative indeterminacy applies equally to 'standard' interpretations based on the assumption of a constant night sky: thus if these are to be upheld, the presence of Taurid phenomena in ancient cosmic cultures, exclusive associations with the Bull or the Ram, has to be explained. In the final analysis, if the constraints imposed by the real astronomical surroundings are not applied, there is little to choose between one school of thought which sees in myth only eclipse dragons, another which sees only alien spacemen, and yet another which sees nothing at all.

One of the most remarkable and widespread beliefs held in early civilizations was that of a god in the sky.[90] Even in the Nile valley, where the Sun was eventually a principle object of worship, there was an earlier celestial religion, older even than the dynasties of the pharaohs. In the Pyramid Texts this early god is shown as the giver of life, of rain, and of 'celestial fire'. Worship of a sky god has been the dominant religious feature of the Indo-European and Semitic peoples from the earliest times. As these groups spread into Asia, India and the eastern Mediterranean, they took their Supreme Being with them.

A popularly held view is that the idea of a Supreme Being developed from the primitive animism of the savage. The story goes that, out of the primitive belief that every tree and breath of wind was alive, there evolved the belief in a multitude of gods, and that out of this polytheism in turn there arose the higher belief in a single Supreme Deity. It is a view which is still popularly held and promulgated. However the result of anthropological researches has been to show that the Supreme Beings of primitive cultures have always been such and have never been 'lower-order' gods or heroes. If anything the evolution has been in the reverse direction, the tendency to create gods being seen, for example, in the proliferation of saints in the Roman Catholic faith. An interesting further finding of this research is that, even in the earliest times, the sky god never existed alone: he was always accompanied by other divine creatures.

Another popular misconception is that the early sky god, like the modern conception of God, was an abstraction. More probably, he was something physically real in the sky, and historians of religion have usually regarded the Supreme Deity as the 'physical sky' itself. Thus behind the Semitic and Indo-European belief in a celestial Supreme Being 'no doubt lay an earlier cult of the sky in which the vault of heaven was deified and associated with transcendental gods and supramundane powers who dwelt in exalted seclusion in the celestial regions'. According to the historian James, the cult of a sky god is very ancient, probably going back at least to the late stone age.

Nor was the original deity a mere weather god. Zeus hurling his thunderbolts from the summit of Mount Olympus is easily seen as a personification of thunder and lightning but there appears to have been an older, celestial Zeus behind this creature. When the Indo-European tribes migrated through southern Russia and along the Danube, some of them settled in Thessaly, near Mount Olympus, around 2000 BC (see chapter 3). They installed their gods on the mountain, under the leadership of Zeus, and in due course Zeus became associated with the weather. But as James points out, he never lost his earlier, celestial connections, the very name coming from a root *div* meaning 'the bright sky'. Their adventures, too, were firmly celestial, Zeus and his followers battling it out with Chronos and his allies in the sky, sending them 'migrating beyond the horizon, heaven knows where'. An illustrative hint of an earlier cometary Zeus is seen in the very old myth in which Zeus was hidden by his mother Rhea from his father Chronos. Because Chronos swallowed his offspring to prevent them dethroning him, Rhea hid Zeus in a cave. There he was suckled by a goat, and from the cavern 'fire flashed forth annually when the blood from the birth of Zeus streamed forth'. Given the cosmological setting of much early myth (see chapter 13), it is difficult to make sense of this story except in terms of an annual meteor stream; but if so, Zeus was, in this myth, a comet. In similar vein, the thunderbolts of the deities were not thunder or lightning. Take, for example, Zeus's response[91] to an imposter Salmoneus who was flinging lighted torches at the sky: according to Apollodorus, 'Zeus struck him with a thunderbolt, and wiped out the city he had founded with all its inhabitants'. Some lightning! Whatever authority one may ascribe to Apollodorus, it is clear from many such remarks by classical writers that the thunderbolts of Zeus were regarded as something quite out of the ordinary.

In the earliest times then, probably going back to the Stone Age, a Supreme Being was worshipped. This being was physically real, celestial, it wielded thunderbolts, and was frequently accompanied by other, lesser celestial deities whose nature we shall investigate in due course.

In *The Cosmic Serpent* we proposed that some of the major celestial deities were originally comets, derived from the break-up of the giant progenitor. As the comets faded from sight, the tales of their adventures, handed down as myth, gradually became meaningless to the tellers and their listeners. However the gods were plainly celestial creatures, and at a later stage of this evolution the names of the chief celestial deities were transferred to the planets, which then, paradoxically, acquired the cometary characteristics of the ancient tales. The bizarre adventures of the celestial Zeus (or Jupiter) for example, did not therefore refer, à la Velikovsky, to the planet now of that name, but to something else.

In fact, as noted previously, the evidence is that names like Jupiter and Saturn were superimposed on the planets in the sky at a relatively late period. The history of the names of the planets can only be traced back with confidence to about 600 BC. Table 4 shows the names which were attached to the planets from that time onwards. Democritus (around 430 BC) did not name the planets and even said he did not know how many there are. The attachment of gods to planets seems to have originated with the Babylonians, who gave two names to the planets, a 'scientific' one and a 'divine' one. It may be no coincidence that a revolution in religious attitudes took place from roughly 600 BC onwards. In the words of one authority:[92]

If we compare the religious attitude of the period before −700 with the period after −300, we detect a tremendous difference. No matter whether we direct our attention to Greece, Egypt or Asia, the differences are everywhere of the same kind. The triumphal advance of cosmic religion and its attendant astrology is an international phenomenon. . . . After 600 BC we can observe the invasion of the Greek world by new ideas and the onset of doubts about the traditional gods. The new ideas induced powerful reactions, such as the condemnation of Anaxagoras for atheism and the execution of Socrates for 'worship of new gods'. [see chapter 6]

The old-time religion, the Babylonian–Assyrian worship of celestial deities, was replaced by new cults and new sets of gods to worship: Mithraism, Zervanism, Orphism flourished in the sixth century BC, only to be themselves supplanted by religions with distinctly monotheistic tendencies and which tended to see the supreme god as an abstraction.

This revolution came after a long period of stability in worship. One might, following custom, attempt to explain it in purely human terms. Independent cities in the Mediterranean – Phoenician, Etruscan but above all Greek – began to move ahead in trade, literacy, technology and democracy, and one might reasonably speculate that the social revolution led the religious one rather than the converse. However, it is not then clear what caused the social revolution in the first place. If, on the other hand, the sky gods were cometary, the universal change in religious attitudes can be simply understood if the original sky gods had merely faded away. There is in fact good evidence that the sky gods did indeed have cometary characteristics. The Babylonian scholar Ilse Fuhr[93] has carried out a study of an ancient Oriental symbol, the so-called omega symbol, and spiral markings associated with it. This symbol is most commonly found on stone carvings and so on dating from the first and second centuries BC but its roots go back much further, probably into prehistory. Unlike other ancient symbols, the appearance of the omega symbol changes considerably from one artefact to another. From its appearance, and its obvious celestial connections, Fuhr has deduced that the omega symbols are representations of comets.

The largest, although not the oldest, group of relics on which the omega symbol is found are the boundary stones, used to delineate territorial boundaries. Marked on these stones are gods, placed as warnings to indicate that the wrath of the gods would descend on the trespasser. There would seem to be nothing abstract or unreal about these gods: symbols on the stones can be recognized as the disc of the Sun and the crescent of the Moon. An example is shown in plate 17, where the omega symbol is placed next to symbols for Anu, Ea and Enlil.

This triad, representing the original great Mesopotamian gods, corresponded at first to male and female deities associated with Heaven (Anu) and Earth (Ea) respectively, the storm god (Enlil) being their joint offspring. In the very earliest representations the members of this triad were usually horned deities, but in later years

Plate 17. (*opposite*) Boundary stone invoking the gods; the omega (comet) symbol appears along with other Babylonian deities such as Anu and Enlil (Cliché des Musées Nationaux, Paris).

Anu took on more the character of the celestial equator whilst Ea and Enlil came to symbolize the zones of the zodiac below and above the equator respectively. This transformation is itself highly suggestive in the context of comets for it reflects rather accurately the way in which cometary dust will be distributed in the sky as a result of orbital precession over a millennium or so, under the perturbing influence of Jupiter. Remarkably, such dust bands have been detected in the sky with the aid of the Infrared Astronomical Satellite IRAS.

The omega symbol is found also on coffin lids, presumably as supplications to the gods, in the company of other sky god symbols; it is found on cylinder seals where kinship between a king on Earth

Plate 18. An Assyrian cylinder seal in the British Museum. Over a holy tree hovers the god Assur with wings and a tail. On each side of the tree there is a king in duplicate. His right hand is raised in a gesture of prayer and his left hand catches a ribbon hanging down from the winged disc. The comet symbol hangs on the end of the ribbon, and one may infer that a divine right to kingship, through identification with the cometary god, is being represented (reproduced by courtesy of the trustees of the British Museum). Similar cometary attachments are associated with Mayan glyphs.

and a cometary deity seems to be represented (plate 18); and it is found also on big limestone blocks from the Anu-Addad temple in Assur, which can be dated to between 1200 and 1300 BC. Here it is found in the company of the swastika. This now notorious symbol appears again, a thousand years later, in China,[94] where it is clearly intended to describe a comet in one of its many manifestations. The origin of the swastika, an ancient (certainly prehistoric) symbol, has always been regarded as mysterious. The facts to be explained are its strange appearance and its universality, for it is found from India to Mexico, from Scandinavia to China. Comets seen in the sky are indeed a natural explanation for this ancient symbol, but not the average transient ghosts one observes now: recurring close encounters with active comets in the prehistoric sky are implied. Fuhr's identification of the swastika as a comet symbol, confirmed by its Chinese representation (plate 19), are thus

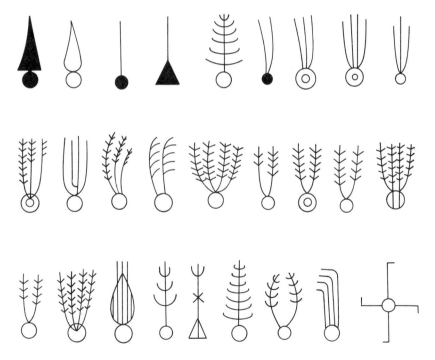

Plate 19. A swastika is perhaps the most surprising symbol to be found within this ancient Chinese classification scheme for comets. The classification is one of several paintings on silk recovered from a Han dynasty tomb dated 168 BC. Examples of multiple tails and fine structure (striations) are also clearly depicted (adapted from the identification chart for the Mawangtui Comet Atlas, 1978).

Plate 20. Nucleus, tail and striations? Possible third millennium cometary motif on the so-called Panorama Stone, Ilkley. Note that the upper ring appears to be a spiral; a megalithic swastika is also found at this locality (photo: Michael J. Stead).

a significant demonstration that such objects were indeed regular travellers in the sky at night.

In fact, one of the impressive aspects of Fuhr's study is her finding that comets may be symbolized by spirals and swastikas, although it must have seemed that these were quite unrealistic representations. But sense can be made of these symbols when it is realized that gas is emitted from a comet nucleus in jets. The gas, spiralling outwards from the rotating nucleus, eventually merges into the surrounding coma and is swept backwards to form the tail. But the spiral form cannot normally be seen without a telescope. For the jets to have been visible to the naked eye there must have been close encounters with bright, active comets. The cup-and-ring markings often found on prehistoric megaliths often have a cometary appearance,[95] and spirals, swastikas and even what appear to be tail striations are to be found amongst them (plate 20). The spiral motif is particularly common in Minoan and Mycenean pottery and frescoes (plate 21).

The omega symbol was not confined to Babylon. Fuhr has identified a 'Heracles knot', an intertwining of omegas sometimes shown as serpents, in many artefacts throughout the near East and seen, for example, in the columned room of the tomb of Ramasses in Thebes. The omega was attached as a sort of label to various Egyptian gods, such as Meschenet and Osiris, for example in a temple of Osiris in Abydos dated around 1200–1300 BC (plate 22); and it conspicuously represents the goddess Hathor through her coiffure (figure 10). Hathor, who was sometimes equated with the body of the sky goddess Nut, was often represented as a cow (plate 5). It was Hathor who overthrew the earlier supreme god Atum after a widespread massacre, Atum having grown old and weak. One of the chief Egyptian deities, she was identified with the Greek goddess of love. Aphrodite, in her earliest forms as the Assyrian Ishtar or the Sumerian Innana, was rather more warlike than her Hellenized version. A Sumerian hymn to Innana[96] gives the flavour:

My father gave me the heavens, gave me the earth,
I am Innana.
Kingship he gave me,
queenship he gave me,
waging of battle he gave me, the attack he gave me,
the floodstorm he gave me, the hurricane he gave me.
The heavens he set as a crown on my head,
the earth he set as sandals on my feet,
a holy robe he set around my body,
a holy sceptre he set in my hand.
The gods are sparrows – I am a falcon;
the Annunaki trundle along – I am a splendid wild cow;
I am father Enlil's splendid wild cow,
his splendid wild cow leading the way!

The pure torch that flares in the sky,
the heavenly light,
shining bright like the day,

Plate 21. Double spiral and omega motifs in Minoan pottery (photo by permission of Princeton University Press). The spiral, the swastika and the omega are found in numerous cultures going back to 2000 BC at least, and may be representations of comets, now defunct, which belonged to the Taurid swarm.

Figure 10. The sky goddess Hathor (line drawing from *Ein Altorientalisches Symbol* by Ilse Fuhr). Associated with the zodiac in ancient Egypt, she is here shown with the distinctive coiffure representing her as a comet. From the Temple of Hathor at Dendera. Such depictions of Hathor go back to at least 1400 BC.

(a)

(b)

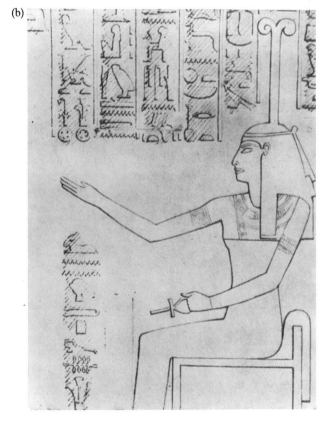

Plate 22. Deities represented as comets in ancient Egypt: (a) Isis and Osiris (defaced by a later pharaoh), as depicted in the Temple of Osiris at Abydos. Fuhr's 'ancient cometary symbol' appears here as the curious twin spiral headgear; (b) The goddess Meschenet, likewise associated with the comet symbol (see Ilse Fuhr, *Ein Altorientalisches Symbol*). The worship of specific gods as comets implies regular recurrence in the sky, consistently with the astronomical evidence for the hierarchic disintegration of a giant progenitor comet into a family of lesser comets during the earliest historical times.

the great queen of heaven Innana,
I will hail.
The holy one, the awesome queen of the Annunaki,
the one revered in heaven and earth,
crowned with great horns, . . .

Innana is usually interpreted by scholars as the planet Venus, and in the absence of the new astronomical information this is no doubt the best that one could have done. However it seems that the settled, spectacular, celestial imagery of the goddess Innana, a morning and evening object 'crowned with great horns' and associated with the omega symbolism of Hathor, is more compatible with the thesis that the goddess was a great comet in a short-period orbit.

We need not go further here. The idea that much imagery in the neolithic and bronze ages is cometary (omega symbols, spirals, cosmic serpents and bulls) is relatively new. Likewise rituals and cults throughout the Near East, in particular the ubiquitous fire-bull rituals of the second millennium BC, have recently been seen as inspired by fireball activity from the Taurus constellation quite independently of the astronomical discoveries mentioned herein.[97]

The three main astronomical legacies from those distant times are celestial myths, calendars with their associated festivals, and megalithic alignments. The field is vast, the attitudes adopted in the face of these legacies are often very rigid and one may well have to wait for the present scholarly generation to die off before a serious re-appraisal of all this material is undertaken in depth. All we shall try to do here is take a cursory look at one of those legacies, and see whether it is better understood from a different point of view: let us take a new look at the old stories.

13
World Ends

IN a pre-literate culture, would such information as the apocalyptic experience of a Tunguska impact be handed down through the generations? The problem, of course, with word-of-mouth communication is that as information is disseminated in space and time, it is likely to become garbled with repetition. In addition, if the information is complex, it becomes difficult to remember. Had twentieth-century communication and transport not been available, it may well be that the Tunguska event would have been obliterated from human memory within a few generations.

And yet, on the other hand, some system for the verbal spread of information, even of a technical nature, must have existed in antiquity: the megalith builders were able to secure stable funding for the various expensive phases of Stonehenge over long periods of time, in addition to aligning their structures on astronomically significant points, without benefit of written memoranda, minutes of meetings and the like. There is evidence that the Indus, Iberian and other traders in pre-history used astronomy to navigate, some reaching possibly as far away as China and Indonesia. The Phoenicians appear to have rounded the Cape of Good Hope, and the Polynesians populated the islands of the Pacific around 1500–1000 BC. All this required a knowledge of the sky, and a degree of reliability in the transmission of technical information over long periods of time.[98]

Some indication of how this was probably achieved comes from the Polynesian navigators of recent times. They had the ability to sail sometimes hundreds of miles without sight of land, and yet find the smallest islands. They did this in part through the reading of ocean wave patterns, and in part through observations of rising and setting stars. The stellar information was given in the form of a fictitious narrative, an explanatory myth, easy to learn and apply. The *Phaenomena*, written by the Greek Aratus (*c*.315–250 BC), comprises star and constellation lore written in poetic form, and may have its roots in the practical knowledge of mariners of the day. It is likely that the constellations themselves, many of which bear little resemblance to the things they are named after, were created out of ancient and universally familiar stories. Information may often have been transmitted, then, in the form of stories, or myth as we would now call it.

Much insight into the nature of these ancient tales has come from studies such as those by de Santillana, von Dechend[99] and others. These studies represent a great advance over earlier interpretations of ancient tales as fertility rites or whatever, or even as the idle fancies of primitive minds. They confirm the very old suspicion that the myths had a *celestial* setting; they were 'vehicles for memorising and transmitting certain kinds of astronomical and cosmological information'. That is, they amount to pre-scientific descriptions of astronomical phenomena. The tales were already ancient by the time they were written down. Encke's Comet, its companions and the associated fireball swarms should indeed loom large in the stories.

There is a preoccupation in myth with long cycles of time. This can be seen in Indian myth especially, world ages periodically coming to an end in celestial catastrophe. They tell us of people who are 'unmistakably identified, yet elusively fluid in outline. They tell of gigantic figures and superhuman events which seem to occupy the whole living space between heaven and earth'. This tendency to describe gods as people extended to sky phenomena generally, geographical features often being placed in the sky. The 'rivers' Oceanus and Eridanus of Greek mythology were conceived in heaven before their transfer to earth. Conversely, 'there are many events, described with appropriate terrestrial imagery, that do not, however, happen on earth'. (The 'earth' in myth is the zodiac, 'dry land' being the band of zodiac north of the celestial equator, the 'underworld' or 'waters below' being the southern half). The inhabitants of this celestial earth are the assorted travellers on the zodiac.

Arising out of her 20-year study of myth, von Dechend made the claim that the ancient tales were describing a slow celestial movement, known as the precession of the equinoxes, several thousand years before its 'official' discovery by Hipparchus in 127 BC. The rotation axis of the Earth is not fixed eternally with respect to the stars: the axis rotates slowly, like the wobble of a spinning top, with a period of 26,000 years. The result is that the celestial pole describes a small circle around the sky with this period, as measured against the fixed background of the stars or the ecliptic, the plane of motion of the Earth. The latter is traced out on the sky by the band of zodiacal constellations along which the Sun, Moon, planets and Encke Comet debris move. The pole star of 4,600 years ago for example, a time of much pyramid and megalith building, was Alpha Draconis. This slow wobble causes the equinoxes, marking the intersections of the celestial equator and ecliptic, to drift slowly westwards through the zodiacal constellations at a rate just perceptible within a human lifetime. During the whole of human history, the vernal equinox has passed through only four of the twelve constellations of the zodiac. The vernal equinox is a calendar marker, the entry of the Sun into it signalling the onset of summer. Slow though it is, precession can be detected within a human lifetime by a slight, progressive shift in the rising point of a star: in the words of one commentator, all one needs is an old tree and belief in the veracity of one's grandfather.

It was also claimed by von Dechend that great astrological significance was attached to precession, the grandest and slowest of celestial movements: it was a great cycle which determined the course of human destiny. Thus there was a Golden Age, from about 6000 to 4000 BC, when the equinox coincided with the Milky Way (being then in Gemini and Sagittarius). This world age came to an end with a great catastrophe. The Phaethon story in which the sun-chariot fell from the sky, setting the world alight describes the end of this age. There followed the Silver Age (4000–2500 BC) with the vernal equinox in Taurus, terminated again by catastrophe described for example in a myth wherein Zeus, during an argument, tilted up a table, causing the Flood of Deucalion. The cosmic myths then, according to von Dechend, have the motifs:

1 Affairs on Earth are controlled by a series of world ages, each of several thousand years' duration.
2 Each world age is controlled and terminated by the passage of the spring equinox from one constellation to another.
3 The opening and closing of these ages is marked by a great catastrophe, in the form of worldwide fire or deluge or both.

It is, of course, catastrophe which catches the eye.

We do not, for our purposes, need to undertake a full critical analysis of the claim that the precession of the equinoxes is being described in these stories. In fact, the orbits of comets and their meteor streams precess at rates which are comparable to those of the equinoxes, with radiants that are very directly seen: thus, if the catastrophe that catches the eye has anything to do with *their* precession, the world ages may well have to be redefined. But for the present, the catastrophe itself is our concern. De Santillana and von Dechend, of course, interpret the catastrophe stories as allegory, a mere vivid way of describing the end of a world age. A few scholars have taken the view that some extraordinary celestial event or events, causing worldwide consternation, is being described in these tales, but for the most part the idea of literal catastrophe has been seen as the hunting ground of the crackpot.

Many other attempts have been made to interpret these stories in allegorical form. One current fashion, for example, is to suppose they represent eclipses of Moon and Sun. On this interpretation the combat is just the chaos in Heaven caused by the eclipse. Indeed Chinese and other myths do describe eclipses in terms of dragons swallowing Sun or Moon, and the term *draconic*, applied to a lunar month connected with the eclipse cycle, reflects this old notion. Some have supposed that the conflict describes the passage of the seasons, some have taken them to represent nothing more than the cycle of day and night, and others have even seen no astronomical content in them, the stories referring to, say, the clash of religions, somehow transferred to the sky. The very variety of explanations indicates that none is completely satisfactory, that a key factor is missing.

One has then the 'allegorical' interpretations of the academics and the 'literal' interpretations of the crackpots. We prefer the latter: the myths are not hiding truth under layers of allegory, they are to be read simply, more or less at face value. Of course the ancient tales may carry allegorical, moral and many other messages, but themes like the creation myths, for example, though coming to us in a variety of clothings, often have a basis so similar that one might reasonably suppose they did indeed start as accounts of out-of-the-ordinary events which were independently observed worldwide – for the stories were told in the Old World and the New. What we must do, clearly, is take a more detailed look at the surviving accounts of celestial catastrophe: plate 23, for example, illustrates a seventeenth-century example of an event in 1667 evidently overlaid with a degree of allegorical interpretation. We must see how they fit the expectations we now have from the new astronomical knowledge.

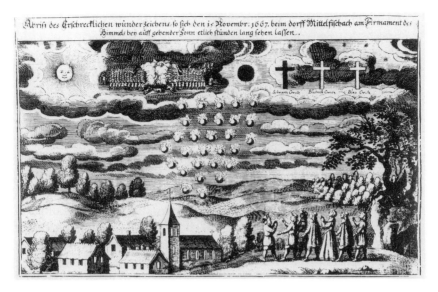

Plate 23. Medieval bolide fall. Seventeenth-century illustration of a swarm of fireballs: stray missiles, so it was believed, resulting from a clash of celestial armies, an idea whose origins may be traced back several thousand years. Note the central presence of a black orb representing the forces of evil (from *The Signs of Heaven* by M. Bischoff).

Specifically, we seek descriptions relating to fireball swarms or Tunguska-like impacts, associated with one or more cometary gods. Zodiacal elements might be sought in the tales: descriptions of disaster coming out of the constellation Taurus would be particularly significant. Finally, such elements should be recorded in myths worldwide.

Certain patterns do indeed appear worldwide in these stories:

1 There was once a supreme sky god; this god often constructed heaven and earth.
2 Later, the 'world' was peopled with giants in the sky.
3 There was a combat involving two major participants, god and dragon, two dragons, or whatever. Often, the outcome of the battle was fire and flood on the Earth below.
4 The major participants were frequently accompanied by hordes of lesser creatures, such as the Annunaki in Babylonian myths.
5 The combat took place in the sky.
6 It was recurrent.

7 The combatants were celestial creatures, giants, noisy, chaotic, winged, serpent-like and they spanned the skies. They might crash to the ground in flames (one curious combatant is the roaring snake which disappears into the ground). The lesser actors often resemble the principals.

We consider that the giants are comets, major splittings from the original progenitor; the hordes are lesser debris; from time to time their orbits bring them into real or apparent conflict with each other and with the real world on the ground; while the world they build is the dust they shed, a battery of luminous rings and orbital tracers. This is not of course to exclude elements of the other propositions which may in part be later additions constructed to give meaning to the tales when orbital evolution was carrying the participants out of harm's way or degassing was losing them altogether: by the time the Greeks recorded the tales, their original meaning may have been completely lost.

In *The Cosmic Serpent*, we pointed out that there is a striking resemblance between the Tunguska impact, a medieval account of a stony meteorite swarm which fell in what is now the USSR (as recorded by a local chronicler) and the two battles of the gods as described by Hesiod in his *Theogony*. The latter, written about 800 BC, describes tales of vastly greater antiquity.

The medieval account, the oldest one of which we are aware, has been quoted by the Russian astronomer Krinov: 'there appeared over the town a dark cloud, and it was dark as the night . . . lightning kept flashing ceaselessly. . . . Even the ground seemed to shake and sway continuously . . . clouds of fire arose and collided with one another, great heat coming from the lightning and thunder'. In table 8 we have listed the main features of the 1908 and 1296 impact events. There is very good agreement between these independent accounts of the two impacts; this gives one confidence that, should these features appear in mythical accounts, we are dealing with a real impact event, whatever the poetic trappings and misunderstandings of later interpreters.

Hesiod's *Theogony* contains a story, probably derived from Babylon or Sumer, that is typical of the combat myths. This is the tale of conflict in heaven between Chronos, aided by the so-called Titans, against Zeus and his five brothers and sisters. These were all children of Mother Earth and Father Heaven. Although represented

Table 8 Impact elements in Hesiod's Theogony

Tunguska (1908 AD)	Velikii Usting (1296 AD)	Hesiod: first conflict (800 BC?)	Hesiod: second conflict (800 BC?)
Blinding ball of fire (darkened sun)		Gleaming brilliance of thunderbolt and lightning (blinding)	Fire from the monster
Thick cloud of dust	Dark cloud	Dust, smoky thunderbolts	Blazing } smoky } thunderbolts
Intense thunder	Intense thunder	Intense thunder, great din	Thundered harshly and strongly
Column of fire	Clouds of fire arose and collided	Immense flame reached upper air	Flame shot out
Blast (flattened forest)		Hot blast, winds	Blast, hurricane winds
Charred trees singed clothes	Great heat from lightning and thunder	Burned forest	Earth caught fire as a result of the awful blast
Ground tremors	Ground tremors	Earthquakes	
	Lightning (ceaseless)	Lightning (thick and fast)	Lightning
		Ocean's streams seethed	Waves raged around shores

as living beings by the Greeks, they were monsters of overwhelming strength. The description of the battle is nothing if not ferocious, for example:

Then Zeus no longer held back his ferocity but now immediately his mind was filled with fury and he showed forth all his strength; at the same time, continually hurling his lightning, he came from heaven and Olympus. Thick and fast, the thunderbolts, with thunder and lightning (note the distinction!), flew from his stout hand and they made a holy flame roll along, as they came in quick succession. The lifegiving earth blazed and crashed all around, and all around immense woods crackled loudly in the fire. The whole land, Ocean's streams, and the unfruitful sea seethed; the hot blast surrounded the earthborn Titans and an immense flame reached the shining upper air. The gleaming brilliance of the thunderbolt and lightning blinded their eyes, strong though they were. An awful heat seized Chaos

and so on.

Later there was another attempt to defeat Zeus, when the Earth gave birth to Typhon, a flaming, winged monster whose head reached the stars, who had a hundred serpents' heads in place of fingers, who was nothing but vipers from the waist down, and whose eyes flashed fire. Typhon was destroyed by Zeus's thunderbolts and fell aflame to the ground. Typhon is explicitly recognized as a comet in classical and early medieval times, being described as such by the astrologer Lydus (AD 490–565), by Plutarch and by Pliny[100] who, 800 years after Hesiod, describes a comet called Typhon having 'spiral-like, glowing red knots'. The spiral motif, as we have seen, could only have been connected with a comet if it were an extraordinary body.

The main features of these battles are also listed in table 8; it is difficult to avoid the conclusion that Hesiod is describing a very large, Tunguska-like impact, large enough to flatten forests. Seasonal or eclipse interpretations do seem a little inadequate, if not silly. But more than this, the impact is clearly associated with its celestial source. The fiery, snake-like monsters in the sky, each with its host of supporters, the recurrence of the cosmic war, the impact motifs, all strongly suggest that our distant ancestors are telling us, through Hesiod, about the very events we seek.

The celestial and cometary imagery are seen quite clearly in other classical authors, as for example in the *Dionysiaka* of Nonnos, written around AD 500. According to Nonnos,[101] the battle began when Zeus hid his celestial weapons in a grotto and they were stolen by Typhoeus. In the course of battle the monster 'lashes the back of the hail-spewing Goat' (*sic*), disturbs the constellations and devastates the earth, and 'uproots trees and forests and rolls them along in front of him'. After the first battle:

night came upon the earth and Helios guided the storm-tossed chariot into the undamaged Atlantic Gates. All the gods wandered fleetingly along the banks of the Nile. . . . Only Zeus still lingered in the high constellation of the Bull. As if driven by the storm, countless leaping stars flashed through the upper atmosphere and, all around, lightning flashes cut the heavens. A comet spread its dishevelled light.

But then, in the morning, Typhoeus 'rose in the shape of a human, encircled the earth with his arms that were a thousand snakes and from his countless heads resounded the howling of wolves, the roaring of lions' etc. After a long and terrible struggle, Zeus's celestial thunderbolts 'smash the crown of the speckled dragon's head, the coiled comet scatters the hair on the head of Typhoeus . . . and the giant . . . now finally plunges to Earth'. The account given by Nonnos, then, although written over a millennium later than Hesiod, contains many elements not in the earlier authors. Certainly, as before, the event is firmly placed in the heavens. But now we have an explicit description of:

1 a fireball swarm, associated with
2 an impact event,
3 emerging from the constellation Taurus, and
4 associated with a large 'coiled' comet.

The statement that the celestial gods wandered fleetingly along the banks of the Nile (only Zeus lingering in the Bull) is of course incomprehensible in terms of terrestrial imagery. However as we have seen the ancient rivers were often cosmic, and the Egyptians connected the Nile and the sky in a very specific way. In this system, discovered early this century by Georges Daressy, the Egyptians mapped the Nile on to the *zodiac*. Nonnos, it appears, is describing the *celestial* Nile, thus:

5 the gods were wandering along the zodiac.

Zeus, the hurler of the thunderbolts, lingered in the constellation of the Bull: one could hardly expect a more explicit description of a fireball swarm emerging from Taurus! A Taurus connection seems to be unique to the catastrophist hypothesis we are proposing (for example it has no eclipse significance), and we shall therefore examine it in more detail.

The orbit of Encke's Comet and its associated material evolves slowly because of the cumulative gravitational disturbances of Jupiter and Saturn. As a result of this precession, the radiant of the Taurid meteors drifts slowly over the millennia, eastwards along the zodiac. Up to two or three thousand years ago the radiant of the southern Taurids was straddling the constellations of Taurus and Aries the Ram; prior to that the meteors were emerging from the Ram. Since we are in almost complete ignorance of the detailed structure of the complex of debris, we can hardly be more precise than say that myths of catastrophe, if based on the giant comet debris, should be based on the Taurid–Arietid region of sky.

The Taurid–Arietid connection is to be found worldwide in the tales of catastrophe, as expected if a universal phenomenon was being observed. For example it is found, in much diluted form, in the Bible (Judges 15):

And he found a new jawbone of an ass, and put forth his hand and took it, and slew a thousand men therewith. And Samson said, with the jawbone of an ass, heaps upon heaps, with the jaw of an ass have I slain a thousand men. . . . And he was sore athirst, and called on the Lord, and said, Thou hast given this great deliverance into the hand of thy servant: and now shall I die for thirst, and fall into the hand of the uncircumcised? But God clave an hollow place that was in the jaw, and there came water thereout; and when he had drunk, his spirit came again, and he revived. . . . And he judged Israel in the days of the Philistines twenty years.

The jawbone of the ass is in the sky: as pointed out by de Santillana and von Dechend, it was the name given by the Babylonians to the Hyades, a bright jawbone-shaped cluster of stars in the constellation Taurus (modern Amazonian Indians call it the jawbone of the ox). In the Babylonian creation epic, older than the sources discussed here, Marduk used the Hyades as a weapon against the celestial monsters. The Samson story, it seems, is a pale survival, the chaos emerging from Taurus reduced to a bizarre fight on earth, the flood connection (p. 181) reduced to a quenching of thirst. The association of the celestial conflict with the Taurus constellation is therefore very old.

In Chinese myth, the world was created by the cosmic serpent Nu

Kua. No sooner was this idyllic world created than the demon king Kung Kung 'rose at the head of a multitude of dragons from the dark waters of the primordial abyss that encircled earth'. After a chaotic battle in which earth and heaven were upset, Kung Kung was defeated and abdicated in favour of his son. His son, Yu Shih, was at times benevolent, at times very destructive; when he appeared on earth with scales of yellow armour and a blue straw hat, accompanied by a one-legged bird, he brought rain; and his throne was in the Hyades. Records show that sacrifices were made to these deities at least as far back as 1100 BC.

In Celtic myth, the monster Gargantua was born from his mother's mouth in the form of a long eel. In this form he corresponds to another Celtic monster called the Tarasque. This latter creature is still celebrated in Tarascon, in the French Pyrenees, in two festivals, according to the anthropologist Huxley,[102] 'one a sober Church festival . . . the other a boisterous occasion when young men splash the girls with water and the monster is wheeled about the town, clashing his toothy jaws at the spectators while his great eyes stare lugubriously'. Variants of this festival survive to the present day throughout the Celtic world. Its significance here is not just the water connection which appears again here: at the same place, but much earlier, Heracles defeated a monster called Tauriscus or Tarusco. In Celtic myth, then, the monster which Heracles fought was a bull. The legend is recorded by the geographer Strabo in the century before Christ but is presumably much older again. The same basic story of celestial catastrophe, then, seems to have been told worldwide, and is at least as old as the second millennium BC.

Grotesque though the birth of Gargantua is, it is paralleled by that of the cosmic serpent Ptah, creator of the earth, who emerged from the mouth of the remote primal Egyptian deity Amen-Khnum. The latter is depicted as a ram-god, and if these tales have any meaning at all, it is at least possible that an earlier emergence of material from the constellation Aries (the Ram) is being described. Zeus is often connected with a ram by the Greek storytellers, as in the combat story where Zeus, pursued by Typhon to the Nile, took the guise of a ram.

It is not to be supposed that the classical authors understood the ultimate origin of the material they were handling. This can be seen, for example, in the decline of Zeus into a mere weather god. Hesiod attempted to give a volcanic flavour to Typhon by having him plunge to Earth under Mount Etna, but the older versions, we know from the oriental sources, have neither weather nor volcanic connections. It is all the more satisfactory, then, that every element one might expect from the astronomical framework has survived the preconceptions of the classical storytellers. Our distant ancestors, it seems, have been telling us in simple language that celestial catastrophe has struck, probably more than once, but the message has been lost through the ravages of time.

Whatever happened was probably experienced over a wide area. Hesiod and Nonnos are far from being original or even principal sources of myth. The Greeks united much local myth which had developed in local communities, but we have seen that the combat myth itself is very much older than Hesiod and occurs over much of the Old World, from Scandinavia to India and China. The conflict between the dragon pair may also be traced back in time to Babylonian myth (Tiamat v Apsu and then Marduk); Indian (Indra v Vitra and then Danu); and so on. Elements of the theme appear in the Bible in the miraculous destruction of Sodom and Gomorrah and the surrounding plain by the raining down of fire from heaven. The story may also be recognized in the Book of Revelation as the battle between Michael and Satan 'the dragon, the primeval serpent'. These terrestrial catastrophes of cosmic origin are discussed in *The Cosmic Serpent* and are only mentioned in passing here (see table 9).

The recurrence of the combat in these stories is interesting: the villains have to be disposed of not once but twice. This could possibly refer to a twin encounter with debris. In figure 9 are shown the nodes of the orbit of Encke's Comet, that is, the two intersections of the comet orbit with the ecliptic plane. Extremely close encounters with the comet are only possible when the nodes intersect the Earth's orbit. For Encke's Comet these intersections occurred around 200 BC and 600 BC, and again around 3900 BC and 3700 BC. As we have seen, if Encke's Comet was active at any of these dates, the intersection would produce the most spectacular celestial phenomena. However, even between nodal intersections fierce annual meteor storms are expected, with peaks in activity recurring at intervals of about fifty years. The dates should not be taken too literally since the encounters may have taken place, not with Encke's Comet, but with a larger progenitor of which Encke is a fragment.

The point to note is that the encounter epochs occur in pairs, separated by a few centuries, that is, within the lifetime of these orally transmitted tales.

The persistence of these stories is as remarkable as their universality: Nonnos is AD 500, Hesiod before 700 BC and the Marduk–Tiamat conflict goes back perhaps to 1500 BC or even earlier. They are in fact recognizably the same basic story. Either the astronomical phenomena recurred over long periods or the verbal transmission of the information was sufficiently robust that the same basic story remained recognizable over timescales of order a millennium. Clearly, the stories were of fundamental importance. Taken together, they amount to a description of perhaps two principal comets in the sky, evidently associated with fireball storms and recurring impacts, some of which seem to have been in the Tunguska class.

Other ancient myth systems, although less well known in the western world than the Greek tales, seem to carry similar celestial messages in different poetical clothing. This can be seen for example in Indian mythology, which is old and complex and formed from a mingling of cultures. Around 2000 BC Aryan-speaking tribes came over the north-west frontier of India from the region of Iran, and settled in the Punjab. The gods they brought with them were already very old and widely distributed, probably going back to the Stone Age. Dealing as we are with gods over four thousand years old we should expect the earliest myths relating to that distant period to include tales of the short-period comets and their celestial debris. The period from these first settlements in the Punjab until about 800–700 BC is referred to as the Vedic Age, during which the Vedas, hymns to the deities, were composed. Other cosmologies, the Hindu, Buddhist and Jain, then arose, but it is especially to the ancient Vedas, with their chants, songs and epic tales, that we should look for evidence of an active sky.

There is evidence in the texts for the existence of gods even older than the Vedic period.[103] These were the Adityas or 'celestials' and included Dyaus-Pitar representing the primeval heavens. However worship of these celestials was already decaying in the earliest Vedic period, and in the later Vedic texts Indra boasts that he has dethroned Varuna, an early great rival of Indra who represents in some way the orderliness of heaven, the regular movement of sun,

Table 9 Possible comet/impact elements in myth and Bible

Phaethon	Typhon and Zeus	Exodus	Book of Revelation	Possible interpretation and comments
Chaos in heaven	cf. table 8			
Sun chariot thrown to ground trailing ashes and glowing dust	cf. table 8		Huge burning star falling from sky	Tunguska-like impact
Phaethon destroyed by a thunderbolt from Zeus	cf. table 8	Storm of freezing hail, lightning and thunderbolts	Hail and fire dropped on the Earth	Likewise the freezing hail is interesting in view of the icy nature of comets. Meteorites are known to have frozen the ground around them on impact
Sun obscured	cf. table 8	Darkness	Sun, Moon, stars darkened by smoke pouring from abyss	Dust from an impact
Earth in flames	cf. table 8		Earth, trees and grass burned with falling hail and fire	Swarm of Tunguska-like impact
Flames followed by flood	Water connection (Strabo etc.)		Burning mountain thrown into sea	Water as well as fire is a widespread motif in dragon myth. (e.g. Leviathan). Precipitation following dust seeding is a speculative possibility

Oppressive fumes from river	Foul-smelling river	Water poisoned by fireball	Open to conjecture but widespread in myth, e.g. Python has a 'noxious interior'
Winged serpents in sky	Winged serpents during subsequent wanderings		Meteorite trails appear serpentine and are frequently depicted as winged serpents in medieval literature
Combatants are winged serpents in the sky; Typhon a blood red, coiled comet (Pliny)	Pillar of cloud by day, fire by night, which moved from front to rear of Israelites	Huge, red, multiple-headed dragon in sky, its tail dragging stars from the sky and dropping them to Earth	Comets e.g. movement of pillar of cloud is that of a comet with a small perihelion (see *Cosmic Serpent*). Revelation dragon must represent a close encounter with a very exceptional body
Recurring battle between dragon pair		Battle between Michael and Satan	Universal theme of oldest myth, apparently referring to a comet pair or orbital precession
Typon as bringer of plagues	Plagues	Plagues	'Sent by gods' in earliest accounts: a large ecological upset is a possibility

stars and so on. We therefore have an early order-versus-chaos
theme. Indra, the principal Aryan (now called Indo-European) deity
was a hammer god like Thor of northern Europe, Zeus of southern
Europe, Ptah of Egypt, P'an Ku of China and Finn-mac-Coul of
Scotland. Indra fought the dragon Vritra in a tale very similar to the
Grecian Zeus–Typhon conflict, one difference being that, whereas
Typhon was buried under Mount Etna in accordance with poetic
preconceptions, Vritra was 'carried to the darksome deeps of
Ocean'. Various groups of demons existed, similar to the gods and
often associated with them, and indeed gods and demons changed
places over the Vedic period. It is not clear whether the demon
groups were different celestial phenomena or different tribal
concepts of the same one. There were the Nagas, serpents from the
waist down, whose king Shesha is identical to Typhon. There were
the 'ugly Vartikas of dreadful sight, having one wing, one eye and one
leg'. These were used in divination: when they 'vomit blood, facing
the sun' this was regarded as an evil omen. It is difficult to see how
non-existent objects, mere poetic fancies, could be used as omens,
and it seems more likely that the Vartikas were real objects, cometary
in nature, the tails streaming away from the Sun and their nuclei
facing it. Imagery consistent with the astronomical expectations is
seen again, for example, in the Titan-like Daityas, who resided in a
moving city Hiranyapura: 'sometimes it sinks below the sea, or
under the earth; at other times it soars across the heavens like the
sun'. The chief deities, Indra and Varuna, have similar celestial
cities, idyllic places like Plato's Atlantis before the fall, or the
Golden Age, or Nu Kua's world before the demons came. One might
seek to dismiss these sky-crossing cities as the products of lurid
imaginations, just as Victorians dismissed myth as the fancies of
primitive minds. But again, if myth is a carrier of cosmic information, they are more consistently understood as real things in the sky.

The slow movement of the equinoxes through the constellations,
and the long evolution of the cometary gods, were the two grand
phenomena of the ancient skies. The world ages were each ended, it
is claimed, by the passage of the equinoxes from one constellation to
another. Whether the precession of the equinoxes was in fact known
in remote antiquity is a debatable question: between noticing the
slow shift of the rising point of a star against an old tree, and
inferring a polar wobble, there is a large theoretical leap. But if we
assume with de Santillana and von Dechend that some village

Einstein took that leap, why should the ancients have described the world ages as ending in catastrophe? Should not the traditions of cosmic swarm be separate from those of equinoctial passage?

The sky has thrown up an interesting coincidence here. The radiant of the Taurids, that is the point on the celestial sphere from which the fireballs seem to come, also moves slowly across the sky, in the opposite direction to the equinox and at a slower rate. The start of the mythic Bronze Age saw the spring equinox passing from Taurus into Aries around 2500 BC, but for the next millennium it also saw the Taurid stream gradually shifting from Aries into Taurus. The constellations themselves for the most part were probably only created around 2600 BC. The Taurids therefore came from out of the spring equinox around the time of the constellation makers. It would be natural, at that time, to link the two great phenomena, the marker ending the world age and the one bringing catastrophe. There is scope here for further study.

The unravelling of the ancient picture will indeed continue to be a complex business, to be approached warily. The point to be made, of course, is that we cannot, as so many mythologists have done (see note 89), try to understand these ancient sky stories while ignoring the state of the sky as it actually was (see table 10). In particular, it is difficult to avoid the signals of catastrophe which were attached to the Taurus and Aries constellations in the remote past.

A comprehensive analysis of the other ancient astronomical legacies, in particular the calendars and alignments, will likewise be a long and complex business, and is not attempted here. Some indication of the way such an analysis might develop can be given, however, by the following example. At present the Taurid meteors appear at night in middle and late November. Allowing for their slow orbital evolution, and for the precession of the equinoxes, one finds that going back in time they appeared about one week earlier per millennium, assuming that they derived mainly from Encke's Comet. In 1000 BC for instance, they appeared in late October/early November, the radiant reaching the meridian at midnight. Again, no great precision can be given since the structure of the Taurid stream is almost unknown. Nevertheless the annual appearance of the meteor stream blazing forth from the sky would make an unmistakable annual event, and one might seek evidence of an annual rite at this date, or even a calendar marker, with fireball overtones.

Table 10 Themes from antiquity: alternative perspectives

Phenomenon	Basic hypothesis: sky dormant in antiquity	Basic hypothesis: sky active in antiquity
Myths of creation	Poetic invention: attempts to understand an overwhelming cosmos in familiar terms	More or less literal description of cometary/zodiacal debris evolution with presence of a major and many minor comets
Temporary worlds and islands of creation, celestial rivers, cosmic enclosures and pillars; creation from mouth of Ram etc.	Poetic invention	Intense zodiacal light and temporary zodiacal structures formed during disintegration of giant comet
Principal celestial gods	Poetic invention: abstractions put in a celestial setting, with weather/planetary motifs added	Real objects, namely short-period regularly visible comets. Weather, planetary or eclipse 'noise' added to cometary 'signal' in course of time
Characteristics of celestial gods; androgynous birth, gradual fading and supplanting by other gods, redness, hairiness, description of Seth-Typhon as comet etc.	Poetic invention	Normal comet appearance and behaviour: splitting, fading etc.

Spirality, omega symbolism etc.	Artistic invention?	Natural representations of comets regularly visible
Myths of catastrophe	Poetic invention: allegorical explanations of eclipses, obliquity of ecliptic, seasonal cycles or whatever	More or less literal descriptions of the Earth's interactions with cometary/zodiacal debris
Thunderbolts hurled by gods	Lightning hurled by a weather god	Impacts associated with a bright short-period comet; 'lightning' interpretation a late addition
Taurus/Aries connection in myth	Poetic invention	Expected
Global nature of calendrical systems, new year starting at culmination of Pleiades or Hyades; fear of world end in October/November	Pastoral or seasonal motivation	Fierce Taurid/Arietid fireball activity
Decline of the gods and celestial myth; rise of rationalism in Greece	'March of progress'	Fading of cometary debris; 'planetization' of myths as original meaning became lost
Dread of sky; sacrifice to deities, especially October/November	Primitive superstition	Knowledge of potential for celestial disaster from Taurid/Arietid phenomena

The natural dividing points of the year would of course be the summer and winter solstices, with the spring and autumn equinoxes midway between them, and there is indeed an abundance of festivals to celebrate these markers. There is also evidence of megalithic alignment towards points on the horizon indicating that a solar calendar was used in neolithic Europe. Curiously, however, later European tribes seem to have divided their year quite differently, this in spite of the fact that, at more northerly latitudes, the contrast between summer and winter is more striking than say at Mediterranean latitudes. Thus the Celts who arrived in Britain from central Europe between 1000–800 BC began their new year with the Samhain. This was one of the four quarter days with which the Celtic immigrants divided up their year (they are still celebrated in Celtic hinterlands such as the Scottish Borders). It fell on 1 November, the date of the Taurids around 1000 BC, was celebrated as a fire festival, and was known as a time when the gods drew near to the earth and the souls of the dead were abroad. Every fire was extinguished until the Pleiades, a conspicuous little star cluster in Taurus, had passed the meridian, at which point new fires were lit and carried by runners to every village. The festival survives as Hallowe'en (see chapter 2) although the bonfires were later transferred to 5 November.

It has been claimed that this calendar has a pastoral basis, marking off the beginning of lambing, the abundance of grass, the exchanging of sheep and cattle at fairs and the slaughter of livestock before winter. Unfortunately for this point of view, as pointed out by Whitlock,[104] pastoral economies came before agricultural ones, whereas the 'pastoral' calendar of the Celts, if such it was, seems to have come after the agricultural calendar. In discussing these ancient calendars, Whitlock also remarks: 'One other small fact slightly worries the author, who was born and bred to agriculture. No farmer that he has ever met has needed a calendar to tell him when to sow and when to reap. Farmers know instinctively and by observation. Why then should so much importance be attached to dates and calendars?' Why indeed, unless the calendar was ritualistic and concerned with the rich variety of Taurid phenomena?

The Chinese still celebrate a Feast of the Dead in November by lighting bonfires and sailing paper boats carrying lighted candles; and at this time in China too, the dead cross the sea to their eternal abode; in Japan, the Feast of Lanterns is celebrated in October and

November; and so on. The returning of the souls of the dead to their homes is very like the Platonic conception of souls carried heavenwards by meteors, and it seems very unlikely that festivals so similar, celebrated worldwide at the same time of year, should have arisen by coincidence. And yet the October/November period has no particular seasonal significance. The festival is as universal, and as old, as the swastika and the tales of celestial catastrophe. If the onset of the Taurid meteor shower was the signal for the start of the New Year, it must have been impressive indeed.

There were two Aztec calendars, a ritualistic one of 260 days and a solar calendar. The origin of the former has never been explained. The least common multiple of 260 and 365 days is 18,980 days or 52 years, and it may be significant too that the 52-year cycle which terminated the Aztec calendars ended in November.[105] At this time there was a festival lasting for 12 days. These acted as 12 leap-days, keeping the calendar adjusted to the seasons. On the midnight of the last day, when the brightest star in Taurus passed overhead (Aldebaran, known to the Aztecs as the star of the fire-making), there was a human sacrifice. All fires were extinguished and then re-lit by torches carrying sacred fire from the temples: the parallel with the Celtic and Asiatic customs is extraordinary. The sacrifices, and other aspects of these festivals, had to do with propitiation and fear, and it seems clear that the source of this fear was the cataract of fire. The alternative is to suppose that these midnight celebrations took place with the Taurids raining overhead unnoticed by priests and faithful alike!

For that matter, certain aspects of Quetzalcoatl, the feathered serpent who represented planet Venus, have a distinctly cometary flavour (plate 24): this is particularly true of its representations in the remote pre-Toltec period. Tales relating to Encke's Comet, like Venus a brilliant morning and evening object, were probably, in the New World as in the Old, eventually transferred to the planet.

The preoccupation with the calendar and cycles of time which characterized Old World societies, the fear of celestial gods which might bring the world to an end if not propitiated, the meticulous observation of the sky, the dominance of theocracy and astronomer-priests, all have their analogues in the ancient civilizations of the west, and all have their internal 'sociological' explanations. In our view, however, the pace was ultimately forced by an external, astronomical trigger.

Plate 24. Crowned with great horns? (cf. plate 9(b); photo by permission of the Museum für Volkerkunde, Berlin.) The feathered serpent Quetzalcoatl on a pre-Toltec stele. The deity symbolizes Venus, as does the disc below. However this depiction, and others, has an obvious cometary flavour, as do catastrophist traditions associated with the god (in some accounts Quetzalcoatl would rain fire from heaven, and throw stones which flattened forests). This is consistent with astronomical evidence that, in early historical times (say before 700 BC), Venus would have been outclassed as a spectacular object by the active comets of the Taurid swarm. It may therefore be that in such representations and tales a memory is preserved of the deity as originally a bright periodic comet of an earlier period.

Of course, one would like to have explicit astronomical records. These are sparse in the extreme. Old Babylonian astronomy of the second millennium BC has left us one astrological and one astronomical text. The astronomical text is the Venus Tablets of Ammizaduga. This is apparently a record of positions of the planet Venus, described as the 'Bright Queen of the Sky'. The Babylonian Venus was therefore the planet as long ago as 1600–1700 BC. This need not be a difficulty for a purely cometary hypothesis since Innana, 'crowned with great horns', would be a brilliant morning and evening star both in her planetary aspect as Venus and in her divine aspect as the giant comet (cf. plate 9(b)): evidence for such duality exists in the dual naming of planets by the Babylonians (table 4).

For the more recent Assyrian period, around 1400–100 BC, the only significant texts are star lists, and the so-called astrolabes. These latter were very common and widely distributed, texts from Assur, Nineveh, Uruk and Babylon having been found, covering almost a thousand-year period. They consist of 36 stars, 3 to a month, covering the 12 months of the year. Evidently, they were calendars, the pre-dawn risings of the stars marking out the course of the year. Oddly, the 'planets' Jupiter, Mars, Mercury and so on are included amongst the star lists. But as calendar markers the planets are useless since they wander along the zodiac and have different positions from one year to the next. Scholarly discussion of this problem is confined to the statement that 'This is very curious, for planets do not appear in a fixed month of the year'. Intense annual meteor streams, recognizably associated with conspicuous comets, would however be natural calendar markers. Associated with the star lists are texts, and in them one reads 'when the stars of Enlil have disappeared, the *great, faint* star which *bisects* the heavens and *stands*, is Marduk: he (the god) changes his position and wanders over the heavens'; or alternatively, 'The *red* star which, when the stars of the night are finished, bisects the heaven and stands there whence the *south wind* comes, this star is the god Marduk' (our italics). As descriptions of the planet Jupiter, these texts appear less than satisfactory, but as descriptions of a great comet in the zodiac with a huge but faint red tail bisecting the sky and associated with a fixed meteor stream (Enlil) from the south, they seem to be more readily understandable. If further textual analysis should confirm these suspicions, one will have a direct identification of Babylonian gods with specific short-period comets most of which, however, will now be degassed and invisible.

So far, the discussion has related only to the last few thousand years, the period of recorded history. The Earth, however, is almost a million times older than this. The story of catastrophe from the sky, occurring within human timescales, must therefore take place in a much larger context. In looking at this larger picture, at the great mass extinctions and geological processes of the past, one may therefore get a better insight into the microcosm which is human history of the last 5,000 years. And in fact, an understanding of the larger picture turns out to be essential for a realistic assessment of future shocks. Let us therefore see to what extent the labyrinth of history, and the bull of heaven, can be united into a coherent model by the thread of science.

III

THE THREAD OF SCIENCE

14

Galactic Imprint

THE Earth is a small planet but a highly active one, much more so than say the Moon, Mercury or Mars. The continents of Earth drift and collide, riding on underlying currents of slow-moving, semi-molten rock below the crust. The viscosity of the material below the crust is high, a billion times that of asphalt say, and it will only move in response to large forces applied steadily for thousands of years. The mobility of the Earth's crust extends to the sea floor: the ocean bed wells up and spreads outwards from long central ridges, before sliding under the continents at the ocean margins. Thus the ocean basins are very young, only a few per cent of the ages of the continents. All these motions are very slow, typically an inch a year; human hair grows faster.

A picture has emerged, then, of a terrestrial crust which is not static, but which is made up of a few continental plates drifting over the Earth's surface. Although the pattern of movement of continents over the globe has really only been worked out for the past 400 million years, less than 10 per cent of the age of the Earth, it appears that in the central regions of North America can be detected the fused remnants of earlier continents and island arcs going back at least two billion years. About 400 million years ago there was, in effect, a single large land mass around the south pole. The mass broke up into a few fragments which, about 200 million years ago, began to move slowly apart. A map of the Earth of, say, 60 million years ago is quite recognizable although, for instance, the Atlantic Ocean was then much narrower. Much of geological history can be understood in terms of the continual drifting and colliding of the giant continental rafts. It is known, for example, that the great mountain-building episodes in Earth history have taken place along the boundaries where the continental plates have been pushed up at the ocean margins, generating earthquakes as they rise and fold, while the sea floor has been subducted back down into the mantle whence it came. The enormous pressure and heat generated at the base of the rising mountains creates molten granite which, forced up along with basalt from the deeper mantle, reaches the surface as lava and volcanic ash. The theory of continental drift was neglected long after its introduction to a sceptical scientific community, and yet (in its modern guise of plate tectonics) it now provides a framework within which many geological processes – mountain building, earthquakes, vulcanism and so on – can be understood. If this were the whole story, there would be no need to look to the heavens for an explanation of events on or under the ground: the Earth can do it all by itself. This has been the traditional view of many geologists, palaeontologists and other Earth scientists.

And yet, within the last few years, it has begun to look as if there is a problem with a purely Earth-centred viewpoint. The problem arises because, if processes on Earth, such as mountain building, were driven only by the random collisions of giant continental rafts, there would be no pattern to them. It seems that this is not so; plate tectonic events recur with some degree of regularity: there are times when the Earth is active and times when it is not. And it turns out that, when we look at the timescales associated with these patterns of activity, a very significant clue emerges.

That there might be some regularity in Earth processes has been suspected for a long time. Over fifty years ago the geologist Arthur Holmes[106] thought that the level of the seas rose and fell in 30-million-year cycles (see figure 11). The changes involved are large, typically 100 metres or more. Holmes also thought that these variations occurred, somehow, in phase with worldwide outbursts of volcanic activity and the like. Holmes went further: superimposed on this cycle, he thought, were even larger events – the creation of great mountain ranges – taking place at about 200-million-year intervals. Such ideas, invoked before the modern understanding of plate tectonics and before rock ages could be accurately dated with radiometric techniques, could be no more than suggestive. In fact, they were largely overlooked in the heat of the plate tectonic revolution. But the ideas of these early workers, so long ignored, bear a remarkable resemblance to new results which are now emerging from researches using improved rock dating methods and the increasing sophistication of geological science.[107]

In 1968, partly from a study of ocean temperatures as revealed by fossil shells, the geologist Dorman suggested that there might be a

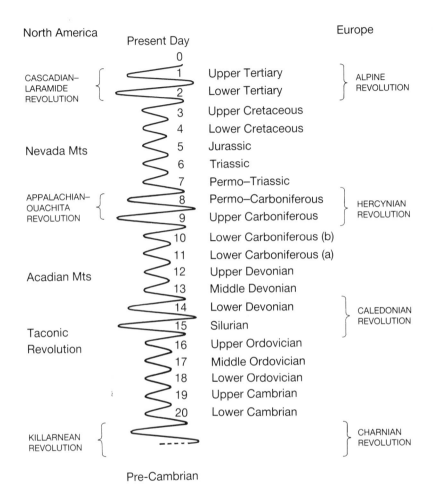

North America

Present Day

Europe

0	
CASCADIAN–LARAMIDE REVOLUTION	1	Upper Tertiary	ALPINE REVOLUTION
	2	Lower Tertiary	
	3	Upper Cretaceous
	4	Lower Cretaceous
Nevada Mts	5	Jurassic
	6	Triassic
	7	Permo–Triassic
APPALACHIAN–OUACHITA REVOLUTION	8	Permo–Carboniferous	HERCYNIAN REVOLUTION
	9	Upper Carboniferous
	10	Lower Carboniferous (b)
	11	Lower Carboniferous (a)
Acadian Mts	12	Upper Devonian
	13	Middle Devonian
	14	Lower Devonian	CALEDONIAN REVOLUTION
	15	Silurian
Taconic Revolution	16	Upper Ordovician
	17	Middle Ordovician
	18	Lower Ordovician
	19	Upper Cambrian
	20	Lower Cambrian
KILLARNEAN REVOLUTION			CHARNIAN REVOLUTION

Pre-Cambrian

Figure 11. Sea level cycles of 30 and 250 million years, claimed by Arthur Holmes in 1927 and correlated by him with periodic geological disturbances (line drawing published by Benn, London). Although this diagram is qualitative and was drawn long before plate-tectonic theory, it anticipates much recent, quantitative research on geological cycles.

30-million-year world temperature cycle (see figure 12). And in 1970, the geologists Hatfield and Camp, from Toledo, argued that catastrophic mass extinctions of life took place at regular intervals. They saw, in the record, seven intense mass extinctions, two of which (65 and 225 million years ago) were exceptionally ferocious. They thought that, overall, mass extinctions happened every 80 or 90 million years, with an exceptionally catastrophic one every 225 to 275

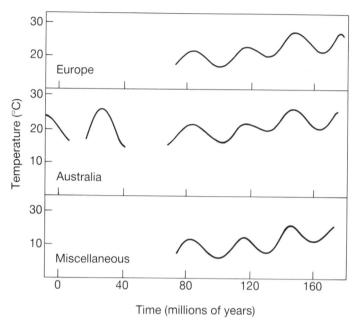

Time (millions of years)

Figure 12. A 30-million-year cycle in global temperature, claimed by Dorman in 1968 using the proportion of 'heavy carbon' in fossil shells as a thermometer. Little weight should be given to the 'cycles' on the left due to the sparseness of the data on which they were based.

million years or so. They thought (correctly) that the Sun completes an orbit round the Galaxy once every 200 to 280 million years, and (wrongly) that the Sun oscillates out of the plane of the Galaxy in a period of 80 to 90 million years. Putting these data together they deduced that the Sun's orbit around the Galaxy must somehow be causing mass extinctions of life on Earth.

In the following year, the Scottish geologist MacIntyre[108] reported an extensive study of the ages of a particular type of rock, known as carbonatite, in Canada. The significance of carbonatite is that it is extruded from the mantle of the Earth during times of high volcanic activity; it therefore acts as a marker for disturbances of the deep interior of the Earth. Any regularity in the production is best found on an ancient stable region such as the Canadian Shield. MacIntyre found the remarkable result shown in figure 13. On this evidence Earth disturbances seem to have occurred, not at random, but at intervals of roughly 230 million years over the last 1,800 million years

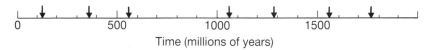

Time (millions of years)

Figure 13. A 230-million-year cycle in the production of carbonatite claimed by the geologist MacIntyre. Carbonatite is extruded in lava outpourings and is a measure of the degree of disturbance of the Earth's outer regions due to the movements of continental plates. This 'grand cycle' is about the length of a galactic year.

at least. There is no significant difference between 230 and 250 million years in the present context – even now geological timescales beyond about 200 million years are not accurate to more than, say, 10 per cent. In effect, MacIntyre's modern study supports the much older and cruder work of Holmes. No internal process is known which could generate convulsions of the Earth at such intervals.

In fact 230 million years is about the time it takes the Sun to orbit once around the centre of the Galaxy. This fact did not go unnoticed by MacIntyre. However he had no suggestion to make as to how something as remote as the Sun's galactic orbit might generate massive outpourings of lava from the interior of the Earth.

Others noted the Holmes timescales in other terrestrial events. In 1973 the geologists Steiner and Grillmair noted that ice ages came and went on the Earth with periods corresponding to the galactic year (say 250 million years) and the Sun's up and down motion through the flat disc of the Galaxy (say 30 million years). Many factors, they considered, must go into the making of an ice age. In order of importance, they listed the distance of the Sun from the centre of the Galaxy, the motion of the Sun out of the plane of the Milky Way, the geometry of the Earth's orbit around the Sun, the position of continents and oceans over the globe, altitude of mountains and plateaux, and unknown causes. In 1977, MacIntyre noted that there had probably been five worldwide geological events over the past 130 million years which were probably caused by sudden changes in the motions of the continental plates. This gives a mean interval between these events of 26 million years, very like the shorter cycle which Holmes thought he had detected. Also in 1977, the American geologists Fischer and Arthur noted that changes in sea level, and mass extinctions of ocean life, seemed to be taking place at 30 million year intervals. Vogt noted that volcanic activity on Earth was not sporadic but that worldwide outbursts occurred at distinct epochs; he, however, thought that the timescale involved was about 50 rather than 30 million years. But still the mystery remained: if the Galaxy was having such a profound effect on the Earth, how?

The first hint (from the geological point of view) at the nature of the cosmic forces involved was published by Seyfert and Sirkin[109] in 1979 (see table 11). These geologists found what appeared to be a direct connection between terrestrial cycles and celestial encounters of a violent kind. Listing the ages of the known impact craters on Earth, they found that these seemed to be bunched, as if impacts were occurring at about 26-million-year intervals. They also found that these bursts seemed to coincide with widespread geological upsets (Holmes again!), caused by sudden movements of the continental plates. The clear implication was that brief but violent bombardments of the Earth were somehow causing terrestrial convulsions. These convulsions are not small affairs: one observes for example a particularly sharp bend in the Hawaii-Emperor chain of Pacific volcanoes, including Hawaii, which occurred 42 million years ago during one of the bombardment episodes. These volcanoes, most of which are submarine, are formed by the drifting of the Pacific plate over a hotspot located underneath the crust of the Earth. The plate has been sharply dislocated and, because they interlock like the pieces of a jigsaw, the entire surface of the Earth must have been affected.

Table 11   Bombardment episodes

Epoch no.	Impact episode (SS 1979)	Impact episode (AM 1984)	Worldwide vulcanisms	Mass extinctions (RS 1984)
1	1—2.5 Myr	—	2 Myr	—
2	15	15	15	11
3	42	40	42	37
4	70	75	70	66
5	100	100	100	91
6	135	130	136	144
7	160	160	159	176
8	180	185	185	193
9	205	210	205	217

SS = Seyfert and Sirkin; AM = Alvarez and Muller; RS = Rampino and Stothers. The 'impact episodes' refer to periods when terrestrial craters were preferentially formed. Note especially that the periodicity 'discovered' in the cratering record in 1984 closely resembles that claimed by Seyfert and Sirkin in 1979. Whether an impact episode is in progress now is left open by these cratering studies, but see table 5.

After many years of neglect, these claims of periodicity are now being rediscovered, using new fossil and other data and sophisticated statistical techniques (table 12). In 1984 Raup and Sepkoski found what appeared to be a 26 million year cycle of mass extinctions in the marine microfossil record, as if the ocean planktonic species were liable to become extinct *en masse* at these intervals. Since then the subject has been controversial though claims for a similar periodicity were made for both the cratering and volcanic records. Some of these data are shown in figure 14. Mathematical techniques exist for picking out any periodic signals which may be hidden away in noisy data and we have applied one such technique to the cratering and volcanic information. There does indeed seem to be a periodicity in the terrestrial record. The period is not, however, 30 million years, as frequently claimed, but 15 million!

A possible reason for this discrepancy becomes apparent when we look at the history of the Earth's magnetic field. The strength and direction of this field at any given epoch in the past can sometimes

Table 12 Galactic periodicities in the geological record

Phenomenon	'Period' (million years)	Author (date of publication)
Climatic and sea-level changes	≈30	Dorman (1968)
	≈32	Leggett et al. (1981)
	≈30	Fischer & Arthur (1977)
	≈30	Shackleton (1989)
Tectonic cycles	≈30	Holmes (1927)
Mass extinctions	≈32	Fischer & Arthur (1977)
	≈26	Raup & Sepkoski (1984)
Geomagnetic reversal frequency	32–34	Negi & Tiwari (1983)
	≈30	Pal & Creer (1986)
	≈15	Mazaud et al. (1983)
	≈15	Creer & Pal (1989)
	≈285	Negi & Tiwari (1983)*
Age of craters	≈27	Seyfert & Sirkin (1979)
	31 ± 1	Rampino & Stothers (1984b)
	≈28	Alvarez & Muller (1984)
Ice ages	≈250	Holmes (1927)
	≈200	Steiner & Grillmair (1973)
Major tectonic events	≈200	Holmes (1927)
	≈230	MacIntyre (1971)
Climatic cycle	≈300	Fischer & Arthur (1977)

* Longer 'periodicity' cycle characterized by a central episode of fixed polarity for 50–100 million years.

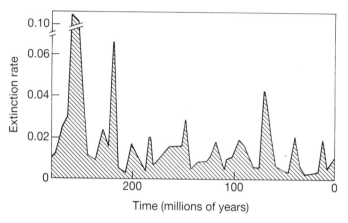

Figure 14. Rate of extinction of families over the last 250 million years, after Raup and Sepkoski (permission D. M. Raup). The arrows represent a 26-million-year cycle. There is no indication of an 'interpulse' between the large peaks but the data may not be at a level where they can be detected. There have not been enough 10-km bolides over the period in question to produce these extinctions, and if due to external forcing, a more continuous interaction between Earth and sky is implied: both the prolonged climatic and prompt impact effects of cometary debris, with galactic modulation, appear to be adequate.

be found, frozen in rock which solidified at the epoch in question. In this way it has been discovered that quite frequently in geological history, the magnetic field of the Earth reverses, north and south magnetic poles interchanging rapidly, within the space of a thousand years or so. The history of the Earth seems to be episodic, long periods of stability when no reversals occur interspersing with episodes of 50–100 million years duration when the magnetic poles flip over at the rate of perhaps five or ten times in a million years. In figure 15 is shown the rate at which the geomagnetic field has been changing over the last 200 million years. From this we can see that, beginning 80 million years back, field reversals have been taking place at an ever-increasing rate. This rate, however, has not been steady: the reversals tend to occur in surges. And in this figure, we see that these surges are periodic, happening every 15 million years or so, the most recent episode having peaked only a few million years in the past.[110]

The reversal pattern, however, is not that of a uniform cycle. Rather, it seems to show weak and strong surges alternating, and another way of describing the record might be to say that there is a

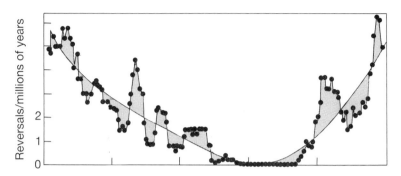

Figure 15. Frequency with which the north and south magnetic poles of the Earth interchange, after K. M. Creer and P. C. Pal. In this more nearly complete geological record a variable 15/30-million-year cycle is clearly seen superimposed on longer-term trends.

strong 30-million-year cycle with a weaker interpulse. This, probably, is why the cycle length detected in the earlier records was 30 million years: as the quality of the data deteriorates, so it rapidly becomes very difficult to detect even the major cycle, and a true 15-million-year weak–strong–weak cycle is misread as one of double the period. The 15-million-year cycle had in fact been thrown up by the mathematical techniques used by several workers examining geological data of high quality, but most of these workers had chosen to ignore the result because it did not fit theoretical preconceptions then in vogue favouring a 30-million-year cycle.

The question arises whether a pattern of this character could somehow be imposed as the Sun orbits around the Galaxy, and the answer is in the affirmative. The Galaxy affects the solar environment in part through tides. As we saw in chapter 9, these tides are negligible on the surface of the Earth or within the planetary system, but they do disturb the orbits of long-period comets, which are only just gravitationally bound to the Sun. The magnitude of the tides will vary with the disposition and mass of galactic material around the Sun, and the flux of comets into the planetary system will vary accordingly. Now it is true that the Sun moves around the Galaxy in a porpoise-like fashion, oscillating up and down through the galactic plane as it orbits the centre. This is likely to yield a 30-million-year pattern of tidal stresses and comet inflow, as galactic material is concentrated towards the plane. However the Sun has also, for the past 100 million years or so, been immersed in a spiral arm known as the Orion arm. It turns out that spiral arms themselves contain substantial amounts of material, enough to create their own tides. Above and below a spiral arm, this tide acts in the opposite direction to that of the galactic disc material. The disc and arm tides cancel each other when the Sun is about half way to the highest point of its orbit, with the result that the comet flux is temporarily halted. The overall result is a basic cycle of 15 rather than 30 million years, with a strong surge when the Sun is near the plane, and a weaker one when the Sun is around the high point of its orbit. The time taken for the Sun to cross a spiral arm is typically 50–100 million years, and these cycles will come and go, episodically, on such timescales.

The origin of the magnetic field of the Earth is not well understood, but it is agreed to be due, in some way, to movements in the fluid core of the Earth. The diameter of this core is about half that of the Earth; it probably encloses an inner, solid core at the very centre. The fluid material of the core, probably a mixture of molten iron and nickel, flows at a rate of about 2 millimetres a minute: a good deal faster than human hair grows. It is likely that the magnetic field at the surface of the Earth is created in some way by patterns in the movement of the molten metal; these patterns must therefore be affected in some way by the Galaxy, presumably through the intermediary of cometary interactions.

Although a reasonably complete understanding of the Earth's magnetic field has not yet been arrived at, its intermittent character is unmistakable. This, coupled with the piecemeal nature of the terrestrial activity, seems to bear out what has occasionally been suspected in the past from the stratified appearance of the geological record, namely the possibility that evolution, even on the largest scales, is not a steady continuous process. Rather, we observe a sequence of unitary events of differing intensity, the intervals between which are ostensibly random though also periodic. As we have already noted, this combination of irregularity and periodicity, with bunching in the record corresponding probably to several galactic cycles over the range 15–250 million years, implies events that are produced in surges which may be attributed to variations in the Solar System environment as the Sun makes its journey through the Galaxy. But the most significant factor so far as evolution is concerned may now be understood as owing more to the effects of recent unitary events than it does to any smooth underlying Darwinian process. Earth history is then in principle a sequence of equilibrium states (involving relatively little evolution) separated by moments of crisis brought on by the arrival in circumterrestrial space of new giant comets. In other words, we deal with a pattern of 'punctuated equilibrium' and a fundamentally catastrophic Earth history.

Such findings do not undermine the popular Darwinian view of evolution but they certainly modify it to a substantial degree. Thus, according to Darwinian theory, we have to imagine some continuous process operating on living species which is capable of introducing small random variations and some other barely perceptible process which allows preferred selections to be made amongst these variations. Such selection, it is supposed, results in that systematic long-term departure from the status quo which we recognize as evolution. The mutation of genes has been identified as the neutral random process whilst the directed process, natural selection so-called, apparently works so as to maximize a species' long-term chance of survival in a given environment – hence, 'survival of the fittest'. The implication is that living or active matter has the capacity to continuously adapt to the environment and reach a state of equilibrium or perfect balance. However, the fossil record is not really like this at all. Whole families, orders or classes suddenly vanish from time to time, each event followed by a period of recovery during which a reduced or modified gene pool, taking species together, is apparently seeking once again to reassert itself over a hostile environment. According to this evidence, evolution may be Darwinian for short spells but the true course of evolution is dominated by the destruction of species during the moments of crisis. Although there remains a sense in which we observe survival of the fittest, life is also subject to an equally significant principle, survival of the luckiest. In recent years therefore, the theory of evolution has been escaping from its Victorian mould.

At the molecular level, the biological process of adaptation is now known to depend on the capacity for self-replication, a property that active or living matter seems to hold in common with inert matter. Inorganic matter for example replicates itself in the form of crystals, whilst organic matter replicates itself in the form of complex structures like DNA. In one case, the outcome is relatively static and robust for a given environment; in the other, it is relatively dynamic and subject to decay. Any difference seems to be one of degree rather than kind, with for example the tobacco mosaic virus having a foot in both camps – in effect a continuum of ever-increasing complexity is assumed: physics into chemistry and chemistry into biology.

On this reckoning therefore, adaptation, or natural selection, is merely the equilibrium-seeking process that we observe taking place on the grand scale in response to minor adjustments to the otherwise invariant process of self-replication at the molecular level. Since the biological record now indicates that the process regularly achieves both new equilibrium and new species in the aftermath of a crisis (or unitary event), then the actual destruction of species as well as the introduction of new components at the molecular level, would seem to be involved in the short periods of change.

Many biological scientists still seek to explain these relatively short periods of change in terms of some kind of instability in the normal evolutionary process which the evidence indicates is generally robust and stable. However it now seems clear from the astronomical perspective that life can no longer be seen as taking place in a closed box: the Earth's environment can no longer just be ignored. Purely internal explanations for periods of rapid change are probably therefore no longer adequate.

It is conceivable that the moments of crisis not only bring conditions that lead to the destruction of species, they may also bring brief additions to the gene pool from outside. The notion that comets are the fundamental source of biologically active material in the universe is clearly an important one, enthusiastically espoused by the astronomers Hoyle and Wickramasinghe[111] in recent years. The idea is very old however: the astronomer William Herschel, for example, who found royal favour during the reign of George III, gave it his support; indeed it could even be said to go back to Plato and in its guise as the source of kingship, even to the beginnings of astronomical theory itself. Whatever the final judgement on this old notion, the history of life on Earth is revealing now that evolution has a cosmic as well as a terrestrial dimension.

Somehow, then, the Galaxy is exerting a decisive control over both geological processes and the evolution of life. If only to assess the current security of *homo sapiens*, we have to look more closely at the mechanisms of control involved. It seems, for example, that the galactic disturbances affect even the core of the Earth, shielded though it is by an overlying mantle of rock 2,800 kilometres thick and with twice the rigidity of steel. How can the Galaxy achieve this feat?

15

Terrestrial Catastrophism

THE long history of the Earth, then, is distinguished by a relatively steady evolution superimposed on which are rapid, sometimes overwhelming changes, whether in the evolution of life, or mountain building, or even disturbances in the core itself. Many geologists over the years have voiced the suspicion that these changes have a galactic periodicity and have speculated that, somehow, the Galaxy is affecting the affairs of Earth. What was lacking in all these speculations was a *mechanism* for inducing the galactic periodicities; and it was not until the late 1970s, with the discovery of the molecular cloud and Apollo asteroid systems, that the mechanism was discovered. There are two main causative links in the chain connecting the Galaxy and the Earth. The first of these is the ebb and flow of the galactic tide which, we have seen, is reflected in the ebb and flow of comets into the inner planetary system. The second link is the existence of giant comets. Rare though they are, it is the giant comets which contain the bulk of the mass which enters the planetary system, and it is their disintegration products which interact most strongly and directly with the Earth. As we have seen, the break-up of a very large comet will produce a wide range of debris, from objects 10 kilometres across, to probably hundreds of kilometre-sized bodies, to temporary swarms of sub-kilometre objects, down to dust, in quantities sufficient to block the path of sunlight. We now ask, how would the Earth machine respond to the influx of such material? Consider, first, an asteroid impact.

The largest known asteroid currently in an Earth-crossing orbit is Hephaistos. It was discovered only in 1978, which gives an indication of the difficulty of finding these objects, whose diameters are only a few kilometres and whose surfaces may be as black as soot. We have seen that this object is ten kilometres in diameter and is in an orbit very like that of the Taurids, but rotated through about 90 degrees. Probably, some tens of thousands of years ago, it split from the giant comet whose debris now circulate along the Taurid track. Hephaistos is a potential collision hazard, although not immediately so. It would strike the Earth at just over 30 kilometres/second, with an impact energy of about 100 million megatons, that is, equivalent to the simultaneous unleashing of about ten thousand full-scale nuclear wars. What would happen?

If Hephaistos struck the ocean it would within seconds generate a cavity in the ocean water and underlying bed about 40 kilometres wide. It would penetrate the crust to a depth of perhaps 20 kilometres, exposing the underlying hot mantle. A column of water, steam, rocks and dust many kilometres high would be hurled into the air, and waves about 5 kilometres high would rise around the impact site. These would drop to a few hundred metres amplitude at 1,000–2,000 kilometres distance but on entering a continental shelf or shallow seas would rear up again into a tsunami several miles high, overturn, and generate intense and prolonged turbulence.

The reefs of the world are found in shallow ocean water, generally less than two hundred feet down. Most of a reef comprises a fine, sandy material loosely bound by the creatures and plants anchored to the underlying limestone framework of coral skeletons. The effect of sudden, violent and prolonged turbulence generated in the seas by a giant impact would be to uproot and disperse the living part of the reef and choke the filter-feeding creatures near the base of the food chain. In addition plankton would suffer through the thick pall of stratospheric dust which would float down through the air and greatly reduce sunlight for at least a few months; if photosynthesis came to a halt, there would be wholesale collapse of food chains. Many modern plankton are already living at the limit of their tolerance of ultraviolet light. Above the atmosphere, ultraviolet light is of lethal intensity. Most of this radiation never reaches the ground, being absorbed by a layer of ozone in the stratosphere. But the creation of nitrous oxides in the fireball would react with the ozone high in the atmosphere, so that when the dust finally settled there would be an ultraviolet deluge and this would itself cause sweeping extinctions. Oceanic ecological systems are complex, particularly reef systems. Because of the lack of seasonal change they are subjected to very little stress and are therefore not adapted to cope with a major upset in the environment. They depend on an abundance of sunlight, which filters down through the water, creating a delicate, stratified system of organisms adapted to the light available at each level. Severe turbulence, lack of sunlight, or high ultraviolet dosage are each lethal to a reef community; and yet these are precisely the consequences of a large impact. It seems that a colliding Hephaistos would wipe out the plankton and reef-building communities worldwide. Presumably, with the base of the food chain undercut, there would be a domino effect throughout the marine kingdom.

Now there is strong fossil evidence for the occurrence of at least five major episodes of marine mass extinction (see figure 16), the largest worldwide in extent and instantaneous to within the resolution of the stratigraphy. Of these, the extinctions in the late Permian 225 million years ago were the most severe; there was a 'mass extinction of truly dramatic proportions, possibly approaching (though of course not reaching) complete extinction of marine life'. To produce the extinction rates observed amongst the classes, orders and families at the Permo-Triassic boundary, something like 90–96 per cent of species may have vanished. The organisms which were destroyed were at all levels of evolutionary development and included the trilobites, five hundred species of plankton, most sponges and echinoderms and, most spectacularly, the corals: thousands of species simply vanished. For ten million years following the event, coral reefs did not appear anywhere in the world. It seems that this event and others like it in the marine record cannot be explained in terms of the normal turnover of species *à la* Darwinian selection; there must have been a worldwide catastrophe which affected life forms *en masse*. After the event, only 2,000–10,000

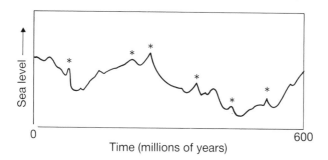

Figure 16. Sea-level variations and episodes of major marine extinctions, according to the geologist A. Hallam, 'Quaternary sea-level changes'. The coincidence of these mass extinctions with major and prolonged sea-level drops is, once again, not easily reconciled with rare huge impacts as a prime cause.

marine species were left. The diversification of life that followed the Permo-Triassic extinction seems to have been in part a matter of chance, some unlikely groups faring well, consistent with random events dominating the survival process: the meek were as likely to inherit the Earth as the powerful. Land extinctions of apparently catastrophic nature were also involved, about 80 per cent of reptile and amphibian families becoming extinct, corresponding to a very high extinction at the level of species.

If Hephaistos were to strike land, a crater perhaps 200 kilometres across would be gouged out within a few minutes, with rims several kilometres high. The basin would flood with lava. An impact in India would flatten forests in Europe, setting them ablaze. Debris thrown out of the crater would range from mountain-sized lumps, themselves formidable missiles, to hot ash, thrown worldwide and adding to the incineration below. Earthquakes would be felt globally and would everywhere be at the top end of intensity scales, with vertical waves many metres high and horizontal ones (e.g. push-and-pull waves) of similar amplitude. These waves would run around the Earth for some hours and would presumably create great devastation. Land creatures would of course suffer also from the cutting out of sunlight for months in equatorial regions and possibly one to three years in polar ones.

The effects of such great impacts then are likely to be felt over at least continental dimensions on land, and worldwide in the oceans. It is not surprising therefore that a connection has often been proposed between past mass extinctions of life on the one hand, and the impact of comets or giant meteorites on the other. Such suggestions could only be speculative because what was lacking was, firstly, a firm astronomical foundation (for example, until recently impact rates were not known even approximately), and secondly, evidence on the ground, say in the form of a large crater. As we have seen, however, a solitary large impact is, from the astronomical point of view, unlikely to be the only agency at work; and indeed, when one looks in more detail at the 'hard evidence', both astronomical and geological, discrepancies in a simple-impact scenario begin to arise.

No extinction has captured the public imagination more, or spawned a greater variety of suggested causes, than the extinction of the dinosaurs 65 million years ago. A thriving and complex community, which had dominated life on Earth for over 140 million years, simply died out. Perhaps some part of the popular fascination with this great extinction arises because it included such memorable creatures as tyrannosaurus rex and diplodocus; or perhaps because it raises the question whether 'it could happen to us'. At any rate, many ideas have been proposed over the years for this mass extinction. One of these ideas, due to the palaeontologist de Laubenfels in 1956 and the geophysicist Urey in 1973, that the dinosaurs were wiped out by a great bolide about 10 kilometres across, was revived in 1980 by a group of scientists at Berkeley, California[112] in order to explain an iridium excess discovered by them at the extinction boundary. The significance of iridium is that it is depleted in the Earth's crust but relatively common in meteorites (say two parts in a million). The presence of this layer was taken to mean that dust had been thrown up from the impact of such a body, settling worldwide. Evidence and hypothesis were announced in a blaze of publicity, the theme was enthusiastically taken up by science journalists, and very soon geochemists and geologists the world over were seeking iridium anomalies and other signs of a great destroying impact.

The Berkeley proposal and the method of its announcement thus stimulated much valuable and detailed research on this important extinction boundary. Others, more traditionally, continued to argue that the extinctions were not instantaneous and that internal processes, dropping sea level and climatic change, were the real explanations for the mass exterminations. As Raup remarked in 1986: 'Some . . . see these developments as heralding major shifts in the way we look at earth history and organic evolution. To others, the past 5 years have seen a science gone mad.' The controversy, at the time of writing, shows little sign of abating, opinions frequently dividing in unexpected directions. For example, the American Earth science community has tended to favour the idea of a great impact while the European one has tended to look askance at the idea; the geochemists have tended to be *pro*, while most palaeontologists have generally been *anti*; and so on. We shall argue shortly that both views are wrong, and that the broad controversy has arisen unnecessarily because proper account has not been taken of the environment of our planet. What is at stake, of course, is an understanding of the history of the Earth. But the argument is not simply an academic game: our assessment of the hazard existing now for mankind depends critically on which of the broad pictures is nearer the truth.

Whether the dinosaur extinction was a single, dramatic event is a highly controversial issue and depends on detailed analyses of fossil remains and their reworking. One school of thought holds that the dinosaurs were an intelligent, complex, thriving community which was overwhelmed by sudden catastrophe. Another holds that the dinosaurs were in decline long before their final extinction. Whether or not the great extinction was instantaneous, it seems to have taken place very rapidly in geological terms, even the non-catastrophe school conceding that the decline in population accelerated rapidly in the last 300,000 years of the Cretaceous period. There were many ocean extinctions too at around this time, including important marine groups such as ammonites, corals and many types of plankton; again, the rapidity of these extinctions is a controversial question, periods of 50 years or less having been proposed.

There are problems with the 'single impact' interpretation of the iridium deposition. In the first place the concentration of the element is too high. An asteroid would excavate several hundred times its own volume of crustal material and this would of course be mixed in with the asteroid material. At many sites, however, the iridium has been diluted by only about twenty times its own volume of material. While high concentrations might occur locally, around the impact site, it is clear that overall there is a serious discrepancy. It has also been argued that in deep-sea cores the iridium is spread through a large thickness of sediment. Burrowing creatures and the tails of fish must disturb the sediment, but this so-called bioturbation may at several sites be insufficient to explain the discrepancy.

Nor does the chemical composition seem to fit the stony meteorite idea too well, the abundance ratios of rare elements such as osmium and rhenium being inconsistent with an origin in such a body. There are enormous and unexplained overabundances in common elements too, such as antimony and arsenic, relative to meteoritic material. And the situation has been complicated by a surprising finding from the eruption in January 1983 of the Hawaiian volcano Kilauea.[113] Airborne particles collected from the volcano were found to have very large concentrations of arsenic, selenium and other elements found in high abundance at the extinction boundary. The particles were also found to be strikingly rich in iridium. Probably, this material has come from the upper mantle of the Earth. This raises the interesting question whether the iridium anomaly might be a gigantic red herring. For this proposition to work, great vulcanisms must have taken place on Earth 65 million years ago, on a scale some tens of millions of times greater than the Hawaiian outburst. There were such eruptions, namely the voluminous out-pourings of lava which gave rise to the Deccan traps in India. The lava from that event contains little iridium; nevertheless, it is interesting to speculate whether, had a volcanic source of iridium been known in 1980, a meteorite impact would have been suggested. It has been suggested that acid rain from this volcanism might have wiped out the dinosaurs.

On the other hand, protagonists of the impact hypothesis can point to an impressive array of evidence. Catastrophic vegetation changes amounting to a 'profound ecological shock' took place across the Cretaceous–Tertiary boundary, at least at sites studied in North America, Europe and Japan.[114] The devastation was sudden and could have been produced by blast, conflagration or sudden freezing of several months' duration. It is also true that the iridium enhancement is sharp at several locations. And in 1985 a most inter-esting discovery was made that very large amounts of soot are present at the extinction boundary in Denmark, Spain and New Zealand.[115] It is difficult to avoid the conclusion reached by the discoverers that wildfires, global in extent, must have taken place during the extinction. On the other hand, not a trace of meteoritic debris, say in the form of stony inclusions in the sediments, has ever been found at the boundary.

How are these discrepancies to be resolved? As we have stated, despite its almost universal adoption by the 'catastrophist' school throughout the 1980s, the stray impact of a 10-kilometre bolide is not, from the astronomical point of view, particularly realistic. Enough is now known about the interplanetary and galactic environ-ments to permit the construction of a realistic astronomical frame-work for terrestrial catastrophism. This framework turns out to be quite different from the simplistic one which has been widely adopted by the new catastrophists:

1 The astronomical environment as we have described it leads one to expect as the prime process, not solitary random impacts but *bombardment episodes*, galactically controlled. These impose a temporal pattern on Earth processes.

2 Within an impact episode, events are dictated by a handful of rare *giant comets*. The disintegration of these bodies liberates copious amounts of dust into meteor streams, the dust spreading out to form dense, temporary zodiacal clouds which may be several hundred

times as massive as the one existing at the present day. The Earth, orbiting within this complex of debris, is subjected to a barrage of large impacts but also experiences large and intermittent inputs of dust into the stratosphere. If the comet is large, and the dust sufficiently fine, sunlight may be greatly dimmed at ground level for as long as the dust cloud or meteor stream persists: this may be hundreds or thousands or tens of thousands of years. If only a hundred thousandth part of a gram enters each square centimetre of the Earth's atmosphere each year, severe climate effects inevitably follow. The Earth becomes blanketed in a white veil. The veil scatters sunlight back into space; but on the other hand it is transparent to infrared radiation, that is heat, from the ground. The Earth can therefore lose heat but cannot gain it. The result is a traumatic drop in the mean temperature of the globe. Over the few million years of an impact episode there could be something like fifty such dustings of the stratosphere.

3 The Earth may run into *swarms* of bodies, each with the potential for numerous blasts and global conflagration, as well as a number of larger Apollo asteroids.

The spectacular prompt effects of a series of impacts, then, will compete with environmental deterioration caused by a series of sharp, climatic coolings. These will lower the level of the seas. A cold Earth transfers water from the oceans; ice caps grow larger and mountain glaciers form. The seas recede from the continental margins, which is more bad news for coral reefs. A great deal of water can be locked up in ice: the present Antarctic ice cap is larger than Europe and two thousand metres deep. In all, ice covers about one-twelfth of the land area of the continents. If it were all to melt, the level of the oceans would rise by 60 metres. The astronomy tells us, then, that an episode of mass extinction of life will often be a complex, prolonged and traumatic affair. This complicated reality will manifest itself in the stratigraphic history of the Earth.

These expectations may be matched against the geological and biological records. In the first place, Earth history does seem to have an episodic character superimposed on a periodic one, the time-scales associated with each fitting the galactic expectations. Second, the high concentrations of iridium found at the dinosaur extinction boundary at several localities, and the absence of bulk meteoritic debris, are hard to explain in terms of a single big bang but easily understood in terms of zodiacal dust as an important provider of the input. Third, there is increasing evidence for a multiplicity of impacts at the dinosaur extinction boundary. This evidence takes the form of various shocked quartz particles,[116] formed in different hypervelocity impacts, and in the precise dating of a small crater, the Manson crater, at the extinction boundary which could not have caused the extinction by itself and which therefore seems to have come in as part of a swarm or shower. Fourth, the predilection for occasional swarms may account for the huge amount of soot at the dinosaur extinction boundary: if laid down instantaneously, about a quarter of the biomass of the Earth was probably consumed in flame. An Earth ablaze is within the capacity of an exceptionally intense swarm to produce, but probably beyond that of even a Hephaistos-class impact.

In 1989, amino acids were found at a well-studied Danish Cretaceous-Tertiary boundary site. These particular compounds were of non-biological origin but of a sort which are not found on Earth. As extraterrestrial markers, therefore, they are probably more significant than say iridium, which might be volcanic. Their occurrence around the dinosaur extinction boundary is therefore important evidence that a major extraterrestrial input was involved. However, it is difficult to see how these compounds, which were laid down over a metre above and below the extinction level, could have been brought in by a 10-kilometre bolide. The fireball in the impact reaches temperatures typically 100,000° C, more than enough to destroy these and indeed any compounds. It seems then that the material probably arrived as comet dust of sub-micron size, trickling down through the atmosphere over some tens of thousands of years. The temperatures reached by such particles during entry are modest and of only a few seconds' duration, enough to permit survival of the compounds.

There are fundamental effects in the response, not only of life on Earth, but also of the Earth itself. In figure 16 the half dozen largest marine extinction events are plotted along with the variations of sea level over the past 570 million years.[117] On each occasion the ocean level has dropped by several hundred metres *before* the mass extinction: it is as if the Earth knew that a huge missile was coming! This has been used as an argument in favour of a purely terrestrial process as causing mass extinctions of species, as has the huge simultaneous vulcanism of 65 million years ago: it is claimed that either or both wiped out the dinosaurs without intervention from the sky.

But of course this uniformitarian argument is unsound. The sea level regressions are indeed surely inexplicable in terms of a 'stray impact' picture;[118] but, as we have seen, it is the wrong picture for the uniformitarians to attack. This illustrates the point that much of the controversy over the impact question has arisen because the protagonists have either misunderstood the nature of the astronomical environment or ignored it altogether.[119] The significance of the correlation between the sea-level regressions and the mass extinctions is that another fundamental terrestrial phenomenon, the level of the oceans, is being affected through stratospheric dusting and the concomitant global cooling.

We can further illustrate this approach by examining the event of about 42 million years ago, which saw the beginnings of the primates, and was connected not only with vulcanisms of the Pacific basin and disturbance of the floor of the Indian Ocean, but also with major changes in the chemistry of the oceans and with a severe cooling of the Earth. The extinctions of that period (impact epoch number 3 of Seyfert and Sirkin) are in phase with the 15- and 30-million-year cycles, as are the dinosaur extinctions.

Detailed studies of the microscopic fossils covering this period have been carried out.[120] These fossils were recovered from deep sea drillings in the Atlantic, Indian and Pacific oceans and should therefore give a global view of the extinctions of the period. About 70 per cent of the plankton became extinct (with a possibly synchronous mass extinction of land mammals). The fossils from the three oceans show that the extinctions did not take place in a single huge catastrophe; rather, they were spread over an interval of several million years. These events are associated with sharp, step-like coolings of the globe, on land as well as in the sea. Clearly, a single impact could not achieve this.

On the other hand, there are indications of cosmic inputs over the time span of the extinctions. About 1–10 billion tons of small glassy stones usually regarded as impact 'splash' material (the North American tektites) cover much of the Earth's surface, but they do not coincide with any of the planktonic extinctions. An excess of the rare element iridium near the boundary was discovered in Caribbean cores. This element is common in meteorites and, we have seen, its presence has usually been taken as an impact signature. There are also some microtektite layers. The iridium layer does not coincide with any of these, having been laid down about 30,000–100,000 years before the nearest of them. This is once again inconsistent with a single random impact but is best explained by the infall of about 50 billion tons of dust (say 50 million tons a year for a thousand years) on to the Earth, followed by the impact of a few-kilometre-diameter fragment from the same meteor stream. The dust input is several hundred times that at the present day. These must have been traumatic times! Once again, then, the overall extinction, lasting some millions of years, is best explained by a bombardment episode of a few million years duration and dominated by the disintegration products of great comets – kilometre-sized asteroids and dust infall from dense meteor streams, the dust especially plunging the Earth into long, cold epochs. The dust intermittently kicked up by the multiple impacts would of course add to the chaos. Overall, the effect is that of a prolonged battering of the environment.

In 1986 the palaeontologist Jablonski[121] reported a detailed study which he had carried out on the extinction pattern of marine bivalves in the Atlantic coastal region. The study covered a 16 million year period which encompassed the great extinction of 65 million years ago. An interesting result of this analysis was the discovery that the pattern of extinction, during the mass event, was not just an intensification of the background extinctions: in particular, species might vanish during a mass extinction for reasons which had no connection with their robustness during normal times. Major environmental change, rather than Darwinian competition, forced the evolutionary pace; and once again, the meek were as likely as the strong to inherit the Earth. The mass extinction was therefore something special, qualitatively different in its effects.

Now, it appears that the semi-regular spacing of such extinctions, studies of which have so far been limited to the last 250 million years, marches in time with the other Earth processes such as the magnetic field reversals, and with the impact episodes as found from the cratering record. Thus, if the Holmes cycles are real, fundamental geological processes are being modulated or even caused by galactic forces.

The evidence for continental drift for example is usually taken to reflect an underlying pattern of continually applied stresses forcing a slow movement of the asthenosphere, the semi-molten rock on which the continental plates ride. But such continuity does not easily account for the intermittent if not cyclic nature of the plate tectonic phenomena. Seyfert and Sirkin proposed that multiple impacts might generate plumes of hot material rising from the mantle, their combined force breaking plates apart and causing them to separate.

The greatest lava outpourings in the history of the Earth have occurred at intervals of about 400 million years, and ocean basins may open and close in about 500-million-year intervals. Over such an interval, the largest single impactor expected is about 25–30 kilometres in diameter. Its impact energy is about 2.5 billion megatons, 25 times that of a Hephaistos collision. Such an energy is about that involved in the formation of the multi-ringed Mare Orientale basin on the Moon. An impact of this magnitude on Europe would fluidize the continent. Brittle fracture of rocks would take place down to about 15 kilometres depth. Along fault lines, rock failure takes place at lower shock pressures and a very large impact would open up cracks over approximately hemispheric dimensions. The opening up of these fissures would produce a fluid magma invading these fissures, and might further drive global vulcanism. The seismic energy from such an impact is best left to the imagination: earthquakes alone might exterminate all vertebrate life on Earth. Plate motions are very likely to be affected by such a disturbance to the underlying asthenosphere. The generation of a crater 500 kilometres wide is likely to involve some vertical crustal motion which will itself greatly perturb the flow of asthenosphere material, at least within a few thousand kilometres. Very large impacts may therefore, as proposed by Seyfert and Sirkin, carry enough energy to affect plate motions.

The South African geologist Hartnady[122] has suggested that a 300-kilometre wide circular depression, about 500 kilometres north-east of Madagascar, may be the site of a great impact of 65 million years ago. He has pointed out that this feature, the Amirante basin, was at the centre of a sudden shift of the mid-ocean ridge at that time, the ridge jumping over 500 kilometres to the north-east. He has also pointed to the existence of a massive submarine slide of sediments, covering 20,000 square kilometres, on the east coast of Africa, as one might expect from the generation of a huge tsunami. Just across the water are the Deccan traps of about the same age, which (we had previously suggested) might have been caused by the triggering of volcanoes from a large impact. It is possible then that a major impact from that period has been identified. It is unlikely, however, to have been the only one, and in fact the shocked quartz particles discovered at the boundary indicate that land impacts also must have been involved.

Smaller impacts generate less disturbance but in the course of an impact episode it is likely that fault lines will be extensively broken by smaller bodies; the disturbances may still be large over plate dimensions and capable of inducing the orogenic and tectonic episodes identified by Holmes, Seyfert and Sirkin and others.

To reverse the Earth's magnetic field, impacts must profoundly disturb the motions in the fluid core of the Earth. An oblique impact will change the rotation of the Earth by a very small amount. In the hours following a large impact, waves will run around the crust and mantle of the Earth, gradually dissipating their energy. Within a week the Earth will have acquired a new direction and period of rotation, differing by say less than one part in 100 million from the previous values. However the liquid core of the Earth does not pick up the new spin immediately and carries on in the old direction. At the interface between the mantle and the core, stresses are set up and there are large disturbances in the fluid velocity. It seems that these impacts may seriously affect the dynamo which generates the Earth's magnetic field.

An indirect but probably even more effective mechanism is the formation of a large ice cap from a sudden cooling. The consequent reduction of sea level, say by 100 metres, increases the equatorial velocity of the Earth by about 0.5 centimetres/second, which is about ten times higher than the flow velocities in the Earth's core. A rapid climatic change from the dusting of the stratosphere would therefore lead to a major disturbance of the flow pattern in the liquid core. Or, the mechanism for both vulcanisms and magnetic field disturbances may be simpler: the mere load shifting as ice caps come and go may set up stresses enough to disturb the Earth in mantle and core, and cause these effects.

It is then possible, in principle, that either an oblique impact or a large climatic disturbance, the latter arising either from impact or direct meteoric input, would upset the geomagnetic dynamo and cause a field reversal.[123] It is unlikely, however, that the sporadic background of ordinary kilometre-sized comets is adequate to force the pace. The tendency to reversal will be most pronounced during a comet shower, when a large comet is shedding its material into the Earth's environment.

The Earth, then, is a huge storehouse of information relating to its continuous interaction with the Galaxy through the agency of very large comets. The variety and complexity of all the physical, chemical and biological processes involved represents but a fraction of the information that may yet come our way to clarify and reinforce our understanding of what is going on. There is no question, therefore, that the current picture of what happens is still incomplete. But the essence of the scientific analysis that is described here is the rigour of the continuing search for the simplest composite picture that meets *all* the relevant data in a straightforward way. One thing stands out however: the importance of very large comets.

However, in arriving at such a conclusion, it is a necessary part of the scientific process that some experts will pursue alternative models, if only to make sure they do not work! Some, for reasons best known to themselves but possibly out of inertia or lack of imagination, will continue to assume the Earth is isolated in space, despite the enormous variety of information to the contrary. Others will select some of the evidence to their liking and invent things in the sky that nobody has seen to meet their particular targets. The nature of our environment is clearly of fundamental interest to mankind and there is always a risk, on account of the subject's importance, that the unwary reader may likewise be captivated by invented schemes that only go a very little way towards explaining everything we see.

For the sake of completeness, we may illustrate this problem by referring to a rather bizarre speculation[124] put forward by two groups of scientists in 1984, one of them associated with the Berkeley stable which had revived the stray meteorite hypothesis only four years earlier. With the rediscovery by Raup and Sepkoski in 1984 of cyclicity in the fossil marine record, and the similar rediscovery of periodicity in the cratering record, it was seen that the stray meteorite hypothesis would have to be abandoned (the long-standing earlier work on these periodicities having until then been ignored). To account for the 'new' cyclicity, a faint companion star in orbit around the Sun was arbitrarily invoked. If this star had a 26-million-year period of revolution around the Sun, and if there is a dense, undiscovered 'inner cloud' of comets which also orbit the

Sun, then every 26 million years the companion, pursuing an eccentric orbit, would dip into this inner cloud and perturb its comets, generating a brief and very intense comet shower. The Berkeley group, with dramatic flair, named the star Nemesis.

The astronomical objections to this proposal are numerous, and in the normal course of events it would undoubtedly have remained an entertaining and obscure speculation. However, once more, the hypothesis was announced in a blaze of publicity, the theme was enthusiastically taken up by science journalists, it became the topic of popular books and television shows, and once more, to many workers in the Earth sciences, unfamiliar with astronomy, it must have seemed that a new and substantial proposal was being made. Certainly the companion-star hypothesis adopts the central mechanism of the galactic one, namely the creation of comet showers through regular comet cloud disturbances. However, the idea that the forcing is caused by a companion star rather than the Galaxy meets with insuperable problems.

In the first place, to have a period of 26–32 million years or so, the binary would be at a mean distance of about 90,000 a.u. (astronomical units) from the Sun, going out to around 180,000 a.u. at its furthest point. Now binary stars are very common in the Galaxy, but pairs as wide as the system postulated are almost completely unknown and are found only amongst very young stars, only a few million years of age. Amongst stars as old as the Sun, the maximum separation of binary components is about 5,000 a.u. The authors of the hypothesis seemed to be unaware that what they were postulating was a system of a type which is, so far as is known from many sky surveys, non-existent.[125] The reason there are no very wide binaries amongst stars of solar age is immediately obvious: we have seen that the long-period comets are stripped away when the Solar System encounters a giant nebula; the same happens to wide binaries. The wider they are, the more weakly they are gravitationally bound to the Sun and the shorter their survival times. It is easily shown that Nemesis would orbit the Sun for only a few hundred million years before being pulled from its orbit by a passing massive nebula and thrown into interstellar space. The probability of its having survived for the 4,500-million-year history of the Solar System is negligible.

It might be that Nemesis was originally in a smaller, less easily dislodged orbit, and has just recently been thrown into its supposed present one. But the Earth periodicities are known to go back at least

200 million years, and it is still very unlikely that the star would remain bound to the Sun over that period. Additionally, although the companion could be given a longer life by supposing it was originally in a tighter orbit, comets would then have been swept up at a much higher rate in the past, say more than a billion years ago. Depending on the structure of the comet cloud, the impact cratering rate from the disturbed comets would have been hundreds or thousands of times higher than the present rate. We have a record of past disturbances in the form of the lunar cratering, and it is clear (figure 17) that comets are most unlikely in the past to have been thrown into the planetary system at these very high rates; indeed the average bombardment rate has not systematically changed for the last 3.9 billion years.

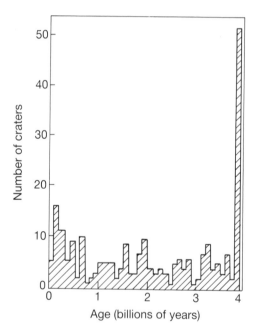

Figure 17. Lunar cratering history, from 3.9 billion years ago to the present (due to Baldwin). The average impact rate has been relatively constant although surges lasting not more than 50–100 million years of uncertain date are also present. A huge reservoir of potential impactors is implied, along with a relatively steady input mechanism. The Galaxy may both replenish and disturb the cloud of comets surrounding the Solar System. The high impact rate 3.9 Byr ago and even higher rates earlier are ascribed to the mopping up of debris left over from the origin of the planetary system, and are apparently unrelated to the cratering history since.

Finally, the companion star is supposed to be roughly at the farthest point of its orbit from the Sun, which would put us half way between bombardment episodes. This is inconsistent with the evidence for a recent disturbance of comet cloud and Earth (table 5). There are other severe problems of a more technical nature, but enough has perhaps been said.

It seems quite extraordinary that a hypothesis so lacking in scientific depth has spawned so much attention, even within the scientific community itself. Normally a new idea, such as this, does not gain approval until it has run the gauntlet of peer review and scientific publication, but in this instance, the attention of the media appears to have been whipped up in advance, more in the style of show business than science, and one can only assume that the normal academic process was somehow distorted by the publicity. Of course, the chief defect of Nemesis is its superfluity, a fact which was concealed by the fanfare of publicity with which the theory was announced. From what is now known of the Galaxy, and the population of small bodies in the Solar System, periodic disturbances of the Earth through bombardment episodes are expected. That these episodes were the cause of great mass extinctions including the dinosaur extinction of 65 million years ago; that the Earth itself would be disturbed through worldwide vulcanisms, magnetic field reversals and so on; and that these global events would be forced with galactic periodicities, had been specifically predicted on astronomical grounds before the iridium discussions or the re-discovery of the terrestrial cycles. Nemesis, like the 'theory of the stray meteorite' which went before it, therefore simply goes beyond the astrophysical requirements at the present time. That is, the Nemesis hypothesis was an unnecessary *ad hoc* response to the terrestrial periodicities, the galactic theory having predicted them in the first place.

The astronomical framework, grounded in celestial observations, is the basis for the theory of terrestrial catastrophism described here. The extraordinary misconception that, on the contrary, proper astronomical theorizing can only follow the terrestrial discoveries, whether iridium or periodicities, has since been put about by some of the Earth science community. Thus the palaeontologist Raup:[126] '*After* Sepkoski and I proposed a 26-million-year stationary extinction periodicity, several astrophysical explanations were proposed';

the geologist Hallam: 'The only hard evidence for disturbance of the Oort (comet) cloud is the iridium anomaly'; and so on. The reader will by now recognize such statements as nonsense, but they are not untypical. Possibly the blazing publicity which accompanied the Berkeley ideas brought about this curious approach to knowledge. More probably the new developments have come as a surprise and what we have observed is a certain lack of reflection as Earth scientists jostle to achieve prominence in what is now widely recognized to be a seminal era in the advance of human knowledge. However that may be, it is in our view essential, if one is to arrive at a true picture, to take account of *all* the relevant evidence: 'hard evidence' in the geologist's sense has to be coupled with some respect for hard astronomical facts as well. Put another way, we do not need a 10-kilometre asteroid to land in our presence to demonstrate the amount of kinetic energy it will release. In particular, the correct picture must explain recent as well as past events in the terrestrial record. Thus the giant comet, and indeed the historical record, are essential elements in the quest for overall truth. It is this inextricable linkage between the very recent and the very remote past which lends urgency to the study: if we get the grand picture wrong, the next set of old bones in the ground could be ours.

16

The Naked Ape

THERE is, of course, a transition between the catastrophism of geology, and that of history. The transition is one of timescale, and it is bridged by mankind. Human beings, in common with monkeys, apes and lemurs, belong to the order of primates. The origin of this order, like much to do with human evolution, is lost in the mists of time and the fog of controversy. It is likely that some time between 40 and 44 million years ago, around the time of the Eocene–Oligocene events, there was a precursor, a creature which was neither monkey nor ape nor lemur, but something in between. Some have identified this creature as amphipithecus, an animal about the size of a monkey whose fossilized jawbones have been found in Burma. If this is correct, then our remote grandparents were fruit-eating, tree-living creatures. Of course, if the ancestral primates originated in southeast Asia, it raises the problem of how their monkey descendants come to be in South America. Possibly they spread into Africa across what is now the Red Sea, and then reached the New World by island-hopping across the Atlantic, which was then much narrower and contained a string of volcanic islands. Another possible route is through northern Asia, across the Bering Straits and down the west coast of North America. Whatever the route, it would have been covered not by purposeful migration but by the random movements of thousands of generations.

At about the time our ancestor appeared, the Earth was seized by a geological convulsion: our roots lie in a bombardment episode. Forty-two million years ago, intense volcanic activity spread throughout the whole Pacific area, from Japan to the Peruvian Andes, in New Zealand and the island arcs and trenches of the Philippines. The sea floor itself, from the Indian Ocean to far northern latitudes, was profoundly disturbed, the floor of the Indian Ocean, for example, beginning to spread rapidly out from a central volcanic ridge. Nor was this great disturbance confined to the Pacific

Ocean. In Africa, there was a sudden uplifting of land in Kenya. The same happened to the western side of North America. And it was a time of intense mountain building which saw, for example, the rise of the Pyrenees in Europe.

Coincident with these great upheavals was a dramatic cooling of the Earth, in the form of a series of rapid and sharp temperature drops. The oceans became cold both at the surface and in deep waters. This was reflected in an accelerated extinction of many important categories of plankton, the extinctions happening in a step-like manner matching the temperature drops. On land, although many parts of the globe remained warm in summer, winters in equatorial regions developed a Siberian harshness. By the time this period had grown to a close, a few million years later, the mammal population had crashed and the forests of the Earth had shrunk: palm trees had vanished from Greenland, and the jungle from England.

As the trees receded, grasslands spread in their place, and grass-eating creatures spread with them. Ancestral horses and camels appeared, and early elephants with long snouts. With the grasseaters came the hunters, primitive cats and dogs. As the climate moderated, the mammals diversified. Many living groups have ancestors going back to that distant time: moles, bats, rabbits, mice, whales, seals, kangaroos and others, about five thousand species in all. It was in East Africa that the evolution of the progenitor humans and apes seems to have progressed. Its climate around 22 million BC was warm and its landscape was a mixture of grassland, woodland and forest. The evolution of the primates can be reconstructed in part from the fossil record, in part from slight biochemical affinities between living species: we carry our remote genealogy in our blood. From such lines of evidence it has been inferred that a growing split between ancestral monkeys on the one hand, and the lineage which led to the orang-utan, the gorilla, the chimpanzee and man on the other, was under way some time after 22 million BC.

Around 15 million BC, the Earth was seized by another geological trauma. This was one of the periodic pulses we have described, the Solar System at this time being at a critical stage in its galactic orbit, and back along the massive Orion spiral arm. Once more, there was a global increase in volcanic activity along with lava out-pourings in many areas around the world. The Alps, Andes and

Taurus mountain ranges rose and folded. There were major deflections in the spreading of the sea floor, and Baha California split from Mexico. This disturbance was of great evolutionary significance: Africa drifted into Europe and Asia, squeezing out the ancient Tethys Sea which had separated them. Thus a bridge was formed, and the ape-like cratures and other mammals migrated across it into Eurasia. The climate in the region straddling the bridge was warm and seasonal, and forests were widespread. It is likely that the great apes developed from an ancestral stock during this time. It is likely also that the very first true hominid, directly ancestral to man, dates from this period. *Ramapithecus* and *Sivapithecus*, whose fossils are found in the Himalayan foothills of India and Pakistan, date from this period. They are, probably, proto-men.

A fog of uncertainty surrounds the evolution of these proto-men for the next ten or more million years. It is known that man-like creatures were walking on two legs around 4 million years ago, their footprints having been uncovered in Tanzania and a half-complete skeleton having been found in Ethiopia. These are the earliest undisputed hominids. They lived in communities, in open savanna. Other than this, little is known of their lifestyle.

Between 2.5 and 2 million years ago, the planet was yet again disturbed. Volcanic ash recovered from ocean sediments obtained during deep sea drilling seems to show a dramatic increase in the rate of explosive activity over the last 2 million years. Certainly on land there was increased volcanism in New Zealand, Japan, the Andes, Mexico, Iceland, the African Rift valley and elsewhere. This trend is illustrated in figure 18 where the number of ash layers measured in various deep-sea drilling sites and lava outpourings around Hawaii, are shown. The oceanic record as a whole goes back only 20 million years, but it is clear from the 60-million-year span covered by the Hawaiian record that there were surges of activity about 40 million BC, 15 million BC and recently (say 2 or 3 million BC).

But now a new element enters into the picture: ice. For a long period of time, going back perhaps 65 million years, the Earth has been cooling. Sixty-five million years ago there were no polar caps, although it is likely that the polar seas froze over in winter. But about 20 million years ago, permanent ice caps began to develop. About 10 million years ago, glaciers began to flow down high mountain tops. And about 2.5 million years ago, ice sheets began to spread over large areas of land in the northern hemisphere. These last 2.5 million

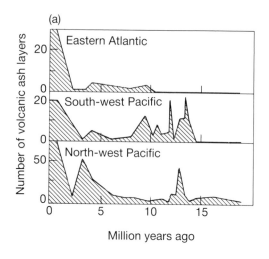

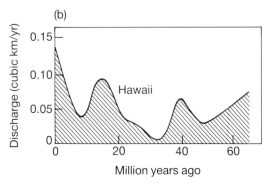

Figure 18. Outpourings of lava from various oceanic sites. Peaks in the output can be seen at around 40 million years ago, some 15–10 million years ago, and within the last few million years. These epochs are consistent with the peaks in the cratering record.

years have seen the onset of what promises to be a long, cold ice epoch on our planet.

At this time also, the African hominids underwent a further evolutionary change. At least two species of proto-men arose. There was a small-brained vegetarian, Australopithecus, who existed for a million years and then, for unknown reasons, became extinct. A creature with a larger brain, *homo habilis*, lasted until just over 1.5 million BC. There is evidence that these primitive African hominids used stone implements.

The first sophisticated toolmakers, probably the direct ancestors of modern man, appear about 1.6 million years BC. Their earliest remains are found in Africa, but by 1 million years BC they were present as far away as China and survived there until at least 230,000 BC. Remarkably, it seems from genetic studies of modern human types that at most a few hundred individuals migrated from Africa and that this single group gave rise to both the Mongoloid and Caucasian races. Be that as it may, a cave near Beijing (the former Peking) was occupied continuously by these toolmakers for nearly a quarter of a million years before the accumulated detritus finally forced Peking Man out. He was able to hunt, he could control if not make fire, and he could cook. And then, less than 200,000 years ago, *homo erectus* was replaced by early forms of modern man. Neanderthal man, the most conspicuous form of these subspecies, was a powerful, muscular creature with strong teeth which may have been used for chewing hides. He was a European, living a hardy life on the fringes of the great ice sheets. Behaviourally, he was quite unlike ourselves. Intellectually, he may well have been our equal.

While these evolutionary trends were under way, the ice epoch came and went in a series of ice ages, each lasting for typically 100,000 years. True modern man appeared 40,000 to 45,000 years ago, probably during a warm period between glaciations. Neanderthal man vanished. But then, 40,000 years ago, the Earth was plunged into another ice age. The temperature fell with astonishing speed, perhaps within a decade, the whole Earth (north and south) becoming about 15°C colder than it is now. The cold reached its peak 20,000 years ago. An ice sheet half a mile thick covered Ireland, Britain, the North Sea, central Europe and northern Russia. Another one covered Greenland and North America down to New York and the Great Lakes. These sheets were thicker to the north, reaching a depth of 1.5 miles over Scandinavia and Canada. Yet somehow a series of migrations took place over the frozen ice sheets and across the land bridge which is now the Bering Straits, so peopling the Americas.

And then, almost as suddenly as it came, the ice age vanished. By around 9,000 BC, it was over; the ice sheets had retreated to Greenland, the huge mountain glaciers had vanished from Scotland and Switzerland, and the climate was suddenly warmer. We can still see the recent signs in the freshly sculpted valleys of northern Europe, the glacial moraines and erratic boulders which strew

Swedish fields, and the fact that Scandinavia is rising, an inch a year, rebounding up from the release of the overlying mass of ice.

The retreat of the ice brought arid, semi-desert conditions to northern Africa, Arabia and the near East. Fertile conditions were now to be found only in elongated strips alongside rivers, or in oases. The human apes could no longer maintain themselves by roaming over wide territories: the domestication of animals and the growth of crops followed. Alongside the great rivers, the need for irrigation and tillage of large areas forced men to co-operate in numbers far larger than a hunting group. Neolithic settlements were replaced by the village and town. Cities then followed, the evident precursors of empires and prospective world domination.

With the departure of the ice, then, the human ape has surged forward, the process of cultural evolution apparently greatly outstripping biological evolution. Whereas in the previous 30,000 years man had hunted in groups, used stone tools and weapons, lived in small settlements, painted and carved, and appears to have had some sort of belief in totemism, in the 10,000 years following the retreat of the ice, civilization has now emerged. Civilization, it seems, is a part of the process enabling the human ape to generate the surplus that supports some of his number as technologists and scientists, architects, artists, writers, philosophers and priests, and it is by the output of these leisured groups that it is often judged. Another measure of civilization is the extent of its control over Nature. By this second yardstick, it seems that the astonishing growth of science and technology over the past three hundred years has put *homo sapiens* in a dominant position, not only within his community of primates but within the animal kingdom as a whole. By now, the human ape can hardly see beyond the cultural surge in which he is enmeshed. He learns to accept leisure and dominance as his birthright. Earth history, human history, cultural history become what they seem, an undeviating progress towards civilization itself.

There is, however, another dimension to the story. For out of the sky, we now learn, there may occasionally appear a cataract of fire, a celestial catastrophe which emerges suddenly to terrorize and destroy. Between cataracts, the human ape is tempted to develop empire and culture, the captivating prelude to dominance and leisure, only to be lulled into a sense of false security, there by failing to see the cosmic forces around him . . .

All of this brings us finally to the question: to what extent are we now secure, and to what extent is our progressive mastery over Nature during the past 5,000 years an illusion, a mere quiet spell before some uncontrollable change overwhelms us?

17

A Risk Assessment

IF the 'stray meteorite' hypothesis, revived by the Berkeley school and uncritically adopted by a large segment of the American Earth science community, should turn out to be correct, the mean interval between devastating impacts is then reassuringly long. Likewise, if the Nemesis proposal is correct, we are half way between comet showers and need not worry about major impacts for another 15 million years. If, however, the galactic hypothesis is broadly right, the story is different. For in that case we are still in the tail end of an impact episode and are living in the wake of the most recent giant comet to enter the realm of the inner planets as a consequence of that episode. This comet, in our picture, has been responsible for a potentially devastating impact in the twentieth century; swarm activity and a near-million-megaton lunar impact in the Middle Ages; a Dark Age over Europe a few centuries earlier; widespread destruction of civilization in at least the second millennium BC, the Biblical Flood at the start of the third; and most traumatic of all, the last ice age, an event which retarded the development of the human species for 20,000 years. Clearly, in assessing the hazards now faced by mankind, we have to get the grand astronomical picture right. Our view, expressed in these pages, is that the scientific case is strongly in favour of the galactic theory.

The astronomical community has so far shown little awareness of the potential human consequences of even a single impact. For example, in a very recent textbook on comets the astronomers Brandt and Chapman state that one expects a Tunguska-like impact about once in every 2,000 years, the devastated area from such being fairly local. They add that 'No one should lose much sleep over this situation even if agitated by uninformed doomsayers.' Leonard Kulik's study of the Tunguska event, on the other hand, led him to a very different perspective:

there is no reason whatever why a similar visitation should not fall at any moment upon a more populous region. Had this meteorite fallen in central Belgium there would have been no living creature left in the whole country. Had it fallen in New York ... all life in the central area of the meteor's impact would have been blotted out instantaneously.

Notwithstanding Kulik's warning, there seems to be a view abroad that the 'responsible' attitude to take towards impact hazards is one of public reassurance, presumably as a counter to uninformed doomsaying. However given what we have now seen of comet Encke's progenitor, one suspects that, behind such attitudes, there may lie uninformed complacency. And, as our 'nuclear error' sketch indicates, there may be a new dimension to the problem which did not exist in Kulik's time. Clearly, it is important neither to exaggerate nor underrate the cosmic hazards revealed in this book, but to discuss them as realistically as possible. We therefore ask: how often do we expect natural missiles to strike the Earth at the present time, singly or in swarms, and how plausible is the scenario with which we (hopefully) alarmed the reader early in this book?

IMPACT RATES

It seems to be a rule of Nature that the larger the body, the rarer it is. Certainly the rule holds for the assorted bodies which belong to the Solar System. Away from town lights, on a clear dark night, and with dark-adapted eyes, shooting stars can be seen every few minutes. These particles are usually only a fraction of a millimetre across. In a few weeks of constant observing, one might see a brilliant fireball, bright enough to light up the landscape as it streaks across the sky. This is usually caused by a fluffy body the size of a pea coming in at about 60,000 miles an hour (25 kilometres a second). At the other end of the scale, as we have seen, we deal with life-extinguishing monsters, bodies tens of kilometres across which unload millions or billions of megatons of energy on to the Earth, and which arrive on geological timescales. But what lies between?

This problem may be approached in a number of ways (see *The Cosmic Serpent*). One way is to look at surviving impact craters on Earth. One must scale up from this by estimating how many craters have gone unrecorded through erosion and so on, and how many missiles failed to reach the ground in the first place. Iron meteorites comprise only a small fraction of all bolides but because they survive passage through the atmosphere intact they are responsible for a disproportionate number of holes in the ground. The trouble with this approach is the extreme incompleteness of the record. Only eight craters probably due to iron meteorites are known, more than 100 metres across; and yet one might reasonably expect five hundred to have been created since the end of the last ice age. Likewise only five iron craters more than a kilometre across are known, whereas the rate at which iron bolides enter the atmosphere leads one to expect fifty to have been generated over the last million years.

The Barringer crater in the Arizona desert is about 1.2 kilometres across and is probably 40,000 or so years old. From the size of this crater it can be calculated that the missile must have had an impact energy in the range 5-15 megatons, and from the debris scattered around it is known also that the missile was an iron meteorite. In fact for the most part *only* meteorites made of iron would reach the ground to produce a crater in this energy range; those made of stone would on the whole produce a shower of little craters; those made of ice, that is cometary fireballs, would break up in the air. The record of falls over the years shows that, amongst the bodies entering the atmosphere with such energies, perhaps about 1 in 70 are irons. Iron meteorites are unlikely to be cometary products and probably form a more steady background of bodies thrown into the inner planetary system from the asteroid belt. This is consistent with the fact that 2 of the 50 known Earth-crossing asteroids have metallic surfaces. Scaling up from an incomplete record with this uncertain conversion factor, one finds that for every impact producing a crater like the one in the Arizona desert, there must have been about 70 which have left no trace; instead of one such impact in 40,000 years we expect one land impact every 600 years. But of course, for every land impact we expect two in the oceans, and these would leave no trace.

This line of argument leads one to expect a collision of several megatons energy to occur somewhere on Earth every 200 years or so. Overall, we reached the conclusion that 'a few dozen sporadic impacts in the tens of megatons, and a few in the 100 to 1,000 megaton range, must have occurred within the past 5,000 years'. For comparison a strategic warhead has typically half a megaton explosive energy. These are average impact rates; they do not take account of the possibility of surges of activity.

Since then, support for these figures has come from a quite

different line of argument. One problem with making these estimates is that the impacts are something like Nature's perfect crime, leaving little or no trace long after the event. The great majority of the bodies carrying up to several hundred megatons of impact energy break up in the air. Those which do reach ground produce shallow craters which soon vanish through erosion and overlaying with sediment. Merely counting one or two holes in the ground and scaling up is not a very accurate procedure. The Moon, however, is subject to the same bombardment history, it has no atmosphere, and very little erosion. A crater once formed is preserved forever or until such time as it is buried under the debris from another crater. The American geologist Shoemaker[127] has carried out a study of small impact craters on the Moon. The results of this work are shown in figure 19. His 'best estimate' rate shows that, over the past 5,000 years, there have been:

- about 17 impacts of at least 12 megatons energy
- about 7 impacts of at least 30 megatons energy, and
- 1 impact of at least 800 megatons energy.

Once again, these are averages. We shall argue presently that the current impact rate may be substantially higher than this. However

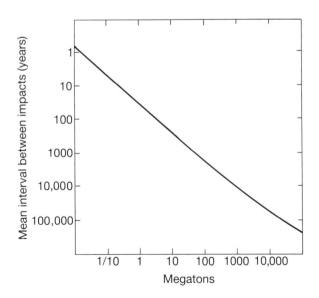

Figure 19. Baseline rate for impacts of various energies on the Earth, after E. M. Shoemaker. Based on the size distribution of small craters on the Moon.

for the moment we shall adopt these as our 'baseline rates', and attempt to evaluate the various cosmic hazards. There are really four scenarios amalgamated in our opening story: a 20-megaton impact is mistaken for a nuclear attack; a 200-megaton impact takes place on an urban area; a 1,000-megaton impact induces a climatic catastrophe; and the Earth encounters a swarm of missiles.

NUCLEAR ERROR

Of all the risks considered here, this is, in principle, the most easily countered: one merely has to be aware of it. Given this knowledge, rational and circumspect men would make no nuclear response. Is the tale then implausible?

There are two main assumptions in the nuclear-error scenario: (1) a Tunguska event, or a swarm of such events, may be interpreted as a low-level nuclear attack; and (2) the likely response to such an attack is an immediate counterstrike. It postulates a vulnerable command which may be wiped out at any moment, so forcing an urgent decision to be made on the basis of a complete misreading of the situation. We shall analyse these assumptions a little more closely.

It might be supposed that with a topic as eschatological as nuclear warfare, every conceivable eventuality has been explored including that of a misinterpreted Tunguska-like impact. Whether this is so or not is not publicly known, the information or lack of it being locked up in, for example, the SIOP books of the United States. However, our discussion of impact rates and swarms in the relevant energy range is based on relatively recent research, and one may ask whether the politicians and soldiers who have evolved the SIOP concept over the past 30 years have fully considered the possibility of such misinterpretation. Some evidence suggesting a lack of awareness of this possibility is the otherwise comprehensive study by the Swiss academic Daniel Frei[128] who was commissioned by the United Nations Institute for Disarmament Research to examine the general question of whether a nuclear war could be started unintentionally. Frei, in his study published in 1982, examines nuclear terrorism, the proverbial 'insane colonel', accidental launch due to human or technical failure, organizational breakdown in a crisis, misperceptions under stress, irrationality of leadership, misunderstandings due to differing perceptions of strategic doctrine, and many other possibilities. But nowhere does he consider the possibility of misunderstood impacts! Thus although we cannot know, we may suspect that in the first chaotic minutes of the crisis, the minds of the commanders will be preoccupied totally with the Bomb, not at all with the Swarm. The risks are enormously compounded by the likelihood that a large impact would threaten the communications systems used for the command and control of nuclear weapons. This arises because of electromagnetic effects which are likely to be caused by the passage of a large moving fireball through the atmosphere and its explosion near the ground. The fireball would interfere with radio and radar transmissions by generating electrons in the ionosphere, create its own radio waves in the fiery tail, and generate a damaging electrical surge, probably thousands of volts per metre, in long cables on the ground. The following effects are tentatively anticipated:

1   Radio propagation from satellites communicating by VHF would be degraded due to the increased electron density, the signal failure lasting some tens of minutes over a region perhaps a thousand miles wide. Those with UHF transmitters would be blacked out for less time over a smaller area. The ground receivers, however, would be subject to a damaging electrical surge.

2   Short-wave radio is the most extensively used system for long-range communication. The radio waves in this range, bouncing successively off ionosphere and ground, may go round the Earth. However an increase in electron density will lead to absorption at these frequencies and so drastically curtail signals. Dumping even a fraction of a megaton at high altitudes may lead to a blackout persisting for several hours over a region more than a thousand miles across.

3   Medium-wave radio is less useful for communications and a significant increase in the D-layer electron density would render them almost useless.

4   Submarine communications are carried out by way of VLF radio waves, which may be 1–10 kilometres long. The D layer acts as a mirror to such waves, which are reflected off its underside and can travel for thousands of miles, guided between D layer and   • ground. Surprisingly, even these may be affected by a high altitude fireball, D-layer changes causing mutual interference

between waves bouncing from the ground and from the sky. Because of their great range, the effect could be important even three thousand miles from a very large fireball.

5 Systems requiring long communication lines (for example, the 2,000 miles between Washington and the North Dakota silos) and solid-state circuits (for example computers) would be vulnerable within perhaps a thousand miles of a large impact.

There is a reasonable prospect, then, that coming in with the devastation of one or more large impacts on a civilized area, there will be widespread and immediate breakdown of radio, radar, telephone, some satellite communications, and computers. A few-hundred-megaton impact on Nevada would affect communications over the whole of the United States, and probably over Canada and Mexico also. It should be emphasized that there are many large uncertainties in predicting the electromagnetic effects of a swarm or even a single large collision. Nevertheless in the present state of knowledge it seems that the first assumption in the nuclear error scenario is not unreasonable: the impact(s) would closely resemble a number of simultaneous ground and air bursts over one or more target areas, in appearance, damage and electromagnetic effects.

Assume then, that in the panic of the moment the National Command Authority believes that nuclear weapons have been exploded on a low-level target in the USA (an assumption more likely to hold during a period of tension). It is likely that, in spite of widespread and immediate disruption, there are surviving channels of communication, and that within minutes of the impact an overwhelming flood of information will start to come in, possibly faster than it can be dealt with. There will be an atmosphere of intense stress if not panic, shock that the explosions could have taken place without warning, and an overriding fear that more bombs may come in at any time. The President or his surrogate will also be subjected to a babel of advice:

The Pentagon advisors will be clamouring to get the President to execute 'their' particular military option, from a host of 'instant' plans drawn up from the information flowing from their pet warning devices. The President will certainly not be left alone in the White House Situation Room for a moment's reflection before he makes the most important decision of his life.[129]

There is in fact considerable doubt amongst experts as to whether a nuclear war could be handled managerially; that is, the 'crisis management' structure would be beyond the capacity of human beings to operate in a real situation. For example, the defence expert Steinbruner[130] considers that great internal conflict would be set up because of the 'inherent tension between the imperative to avoid war and that of controlling the circumstances of initiation if war cannot be avoided'. We have in fact made this a central decision-making issue in our story.

Emotional stress is a particular hazard. There have been many occasions when leaders have buckled under the strain at critical moments in war, and no more stressful situation can be imagined than the one postulated here. Many studies have shown that, beyond a certain level, the ability of individuals to think through a situation deteriorates. Misperception may arise through information overflow. During fast-moving events those at the centre of decisions are overwhelmed by floods of reports compounded of 'conjecture, knowledge, hope and worry'. In coping with the flood there is a tendency to 'omission, error, queuing, and filtering'. Under stress reality becomes distorted; new evidence is looked at with the eye of preconception; adversaries are seen in stereotyped form; there is a tendency to jump to conclusions; a loss of perspective; a searching for scapegoats; exaggerating the positive sides of a decision and minimizing the negative; and so on – in all at least seventeen psychological effects of severe stress have been identified.

Stress affects groups as well as individuals. The need for quick action contracts the group to a very few, enhancing the problem of information overload. 'Groupthink' emerges strongly in a crisis – there is no time for dissenters and experts outside the small circle; there is an illusion of invulnerability; collective rationalization; a failure to look at alternatives; a failure to rethink in the light of new data; a failure to balance risks properly. What Frei and Christian call 'the dangers of cognitive distortion' are exacerbated by the complexity and sophistication of modern strategic doctrines. Like Steinbruner, they question whether the decision makers, 'in the hectic situation of a crisis emergency, will still be capable of deciding and acting fully according to the requirements of the complex logic of strategy ... one may ask whether they are also in a position to understand all the contingency plans and flexible and varied options offered by today's nuclear arsenal'.

None of this gives one boundless confidence that good decisions will be made in the chaos of the moment. For example the apparently idiotic suggestion made by one of our fictional characters, that pointless targets were chosen so as not to justify the risk of retaliation and yet act as a 'keep out' warning in relation to some major military incursion, is in fact a well-recognized war scenario, the 'called bluff', discussed by Frei and others. In the stressful, countdown atmosphere of the Situation Room, and lacking recognition of the astronomical alternative, the suggestion might not seem so strange.

If there is time to ponder, and if unacceptable risks are not imminent, a 'surprise attack' ought not to lead to accidental war. If however the issue at stake is seen to be a vital matter such as the imminent loss of nuclear control, command and communications, then:

Although many authors assume a particularly cautious and sober behaviour in this type of situation, the findings of comparative research on international crises . . . do not give many reasons for optimism. There is always a grave risk of suboptimal behaviour. What this means in practice is easy to realise: miscalculation, false perceptions, and all sorts of inappropriate reactions.

The second assumption of the story, then, appears to be defensible.

Would a limited nuclear exchange remain limited? This question has been the subject of much discussion and literature, the upshot of which is that nobody knows. On the one hand historical precedent might lead one to suppose that in a no-win situation even a modicum of rationality would immediately bring a mistaken exchange to a halt. On the other hand, once triggered, the nuclear-exchange situation seems to be uniquely unstable. Certainly the idea of limited nuclear war has always been rejected in Soviet circles, the late L. Brezhnev, for example, stating that 'if nuclear war breaks out in Europe or elsewhere, it will necessarily and unavoidably become universal'. Pringle and Arkin, in their study of SIOP (single integrated operational plan), conclude that 'In the present SIOP only one choice before the President is still convincing. That choice is the Major Attack Option'. And Steinbruner considers that if war appeared unavoidable, military commanders would put irresistible pressure on political leaders to get missiles away whatever prior policy may have been. It seems that the Secretary's 'controlled response' carries the risk of uncontrolled escalation.

Between them, the USA and USSR occupy about one-seventeenth of the surface area of the Earth. The number of targets which could be termed 'strategic' in each territory is probably several thousand. Siberia, for example, is now dotted with new cities and industrial complexes. If a single missile of Tunguska proportions or greater now brings in with it the risk of 'nuclear error', then a story like the one above may be enacted once every three or four thousand years. For a male in western society, the probability of this event happening within his lifetime is not too different from say the probability that he will die of lung cancer. It is interesting to compare this with the other nuclear risks which Frei and Christian, rather boldly, tried to quantify. They estimated the probabilities that a major nuclear war would be launched within five years, under 'normal' conditions, due to the following factors:

- technical failure   1 in 10,000
- the urge to pre-empt   1 in 10,000
- nuclear proliferation   1 in 10,000
- human failure   3 in 10,000

To these we must now add:

- misinterpreted impact   10–20 in 10,000

assuming that, given the impact(s), the risk of misinterpretation and counterstrike is high. The risk that nuclear war would be accidentally triggered through one or more large impacts would seem on this evidence to be at the level of about 2 per cent within a lifetime.

There are, of course, very large uncertainties in all these estimates, and the object of the scenario is not to put forward a particular view so much as to draw attention to what seems to be a largely unrecognized hazard. It may be more likely than not that a huge impact or even a swarm would immediately be recognized as such. And even if not, perhaps calm and rational minds would prevail in the minutes following catastrophe. But we should at least be aware of the risk. As Frei and Christian put it, there are 'monumental gaps of knowledge which have to be filled by speculation. Yet, as this speculation pertains to deadly risks, it would certainly be unwise to dismiss the more "pessimistic" conclusions out of hand. Dangers which are conceivable in principle are also possible in practice'.

Once the problem is recognized, the cure is, in principle, achievable by a mere technical fix, an extra layer inserted into the decision-making structure. Technical solutions for the larger impact hazards, however, are less easily found.

URBAN IMPACT

Let us adopt the 'random-impact' picture in which the possibility of bunched impacts is neglected, and take the conservative 'baseline' rate of one impact on Earth, of 10 megatons energy or greater, per 200 years. There is then an expectation of about 25 such impacts over 5,000 years, about 7 of which are on land. Cities of more than a million inhabitants presently occupy about 500,000 square kilometres (surprisingly, missile silo fields take up about half this area). A direct hit on a city will occur, then, with only 2.5 per cent probability over 5,000 years or, within a human lifetime, with only about 0.03 per cent probability. These are reassuringly small figures.

However, inhabited areas occupy much more space than cities alone. The present world population density is only about 10 per square kilometre but of course varies from nil for the ocean surface, to 80 for western Europe and about 400 per square kilometre for say the UK and Netherlands. Of course, the city concentrations are much higher, Manhattan having about 25,000 people to the square kilometre for example. Furthermore these population densities are increasing rapidly. Whereas 2,000 years ago the human population of the Earth was about 250 million, by 1963 it was three billion, fifteen years later (1978) it was 4.1 billion, and predictions for AD 2000 range from 4.5 to 7.6 billion (the United Nations official forecast being 6.7 billion).

For an impact on a fairly average town-and-country population numbering 100 inhabitants to the square kilometre, about a million people would be within 30 miles of ground zero. It is likely that a Tunguska impact would promptly kill most of them: the damage done to central Belgium as described in the opening scenario was based on simple extrapolation of nuclear weapons studies. With the figures adopted, such an urban impact is expected two or three times within 5,000 years or, within a human lifetime, with about 3 per cent probability. These are still reassuringly small rates although we shall see that they may have to be revised substantially upwards in the light of various factors. However, when we come to impacts with energies around a thousand megatons, a different story emerges.

COSMIC SWARMS

The Earth and lunar cratering rates are in rough agreement. However, these rates are *averages*; they tell us nothing about the present-day impact hazard, and they might conceal brief periods when the impact rate was enhanced, perhaps enormously so. Scaling up from a few Earth craters effectively programs swarms out of the calculation. The situation for the lunar cratering is even worse, since the computed ,yield is an average taken over the age of the lunar surface, about 4 billion years.

There are in fact indications that the population of interplanetary bodies generally is currently much higher than the cratering record indicates. For example the arrival rate of small bodies which burn up in the atmosphere is 10–100 times higher than one would expect from the crater production rate. This estimate is possible because over the last twenty years Continent-wide networks of all-sky cameras, covering almost a million square miles of sky, have been recording meteoric events. One of the largest fireballs measured over this period had half a megaton of kinetic energy and burned up over the South Atlantic. This is about 25 times the explosive energy of the Hiroshima bomb and illustrates that the upper atmosphere shields us from most impacts in this 'low energy' range. The protection does not extend above a few megatons. This high fireball rate is consistent with the general overabundance of meteoroids in the inner planetary system, as manifested, for example, by the Stohl stream, and with the fact that Apollo asteroids appear to be several, possibly many, times overabundant in relation to the lunar cratering.

However, even the modern fireball rate tells us nothing about the prospect of a swarm of Tunguska-like missiles: one might as well try to infer the occurrence of a fireball storm from an average night's meteor observing. Over and above any temporary enhancement of the 'baseline' impact probability, say over the last few thousand years, there is the evidence for strong bunching of impacts, such as the 1975 lunar-impact event, when for a few days the Moon was bombarded at several hundred times the background rate, and the occasional mention of daytime fireball swarms by the Chinese chroniclers. To proceed, we have to consider not only what factors might enhance the impact hazard overall but also what might cause impacts to bunch briefly in time. That is, we need to apply a realistic astronomical model. It is at this stage that an understanding of our Galactic and Solar System environments becomes crucial.

From the Sun's current position in the Galaxy it is clear that the impact rate at the present time must be exceptionally high. If Earth history is episodic due to passage through spiral arms, then the Sun's position at the inner edge of the Orion spiral arm ensures that we are currently in an active phase. Further, the Solar System has just passed through the plane of the Galaxy, where the tidal stresses acting on the comet cloud are at their maximum; the comet flux is therefore near a strong peak of its galactic cycle. It has also recently passed through Gould's Belt and is therefore undergoing an exceptional tidal stress due to a recent passage through an old, disintegrating molecular cloud. Specifically, although the detailed modelling is uncertain, there is good evidence that only a few million years ago the Sun passed through the Scorpio–Centaurus association, a group of young, hot blue stars and massive nebulae belonging to the Gould's Belt complex. This encounter must have created a sharp impact episode, within which we are still immersed, most of the flooding of the Solar System taking place within a few million years of the encounter.

Evidently then, the Solar System is currently in a very noisy galactic neighbourhood: we are in an impact episode now, just past the peak of a strong galactic cycle, whose effect has been enhanced by a recent disturbance due to passage through a massive nebula. Conservatively speaking, it is likely that the impact rate over the past two or three million years has been several times higher than one might expect just by looking at the lunar craters. This is consistent with the evidence on the ground in the form of recent impact craters and recently disturbed geology. It is also consistent with the fact that the current population of Earth-crossing asteroids would yield too high an impact rate on the Moon to be consistent with the observed number of craters.

Even within a bombardment episode however, giant comets are around for say only 10 per cent of the time. Overall, a giant comet may therefore orbit within the inner Solar System for only a few per cent or so of the time. At one extreme then, if these great bodies are the prime suppliers of missiles, the impact rate averaged over the last

50,000 years say could be as much as a hundred times the rate averaged over the history of the Earth.

If that situation pertains now a number of interesting consequences follow. Adopting our baseline rate, an impact like the Giordano Bruno one would take place on Earth once in 200,000 years; within the timescale of civilization, there is then only a 2.5 per cent probability that such a collision would have occurred on Earth or Moon. However with a rate momentarily 100 times higher, the probability becomes 92 per cent. Far from being a remote statistical fluke, the lunar impact of AD 1178 becomes a not unreasonable expectation. This enormously enhanced risk, persisting for only a few tens of thousands of years, would not greatly affect the long-term average impact rate as deduced from the lunar cratering.

It also follows that Tunguska-like impacts occurring at a hundred times their long-term average rate would yield a current average rate of one impact every other year. These would not in fact occur every other year, but would rather be confined to those brief periods when the Earth intersected the orbit of the comet; further, should the creation of boulders take place by the disintegration of large bodies, Tunguska-sized boulders would then exist, temporarily, in swarms, and impacts would occur primarily when the Earth encountered a swarm. The hazard would be greatest at one or two particular times of the year, for a period of a few centuries, interspersed with calm periods lasting for a millennium or so. Thus there would be many years when nothing happened, and one year when the Earth was intensely bombarded over a few days or hours.

The detailed profile of swarm encounters is determined by factors such as the rate at which the orbits of the debris precess in space, the thickness of the 'meteor' stream, the mass distribution of boulders and so on. Suppose, for example, that the impact rate is currently ten times above the lunar average, and is concentrated in brief bursts when Earth and swarm intersect. Then one finds that a full-blown cosmic swarm would be expected about once in 5,000 years and would comprise an impact of at least 10,000 megatons energy, half a dozen impacts of at least 1,000 megatons, 40 of at least 100 megatons and 250 impacts of at least 10 megatons energy. Our opening scenario is a good deal more modest than this.

That such encounters could happen is not to say that they would happen: the argument so far is one of consistency rather than proof. What physical factors might in reality cause boulders to concentrate in swarms rather than, say, be spread uniformly throughout the inner planetary system? A picture is emerging of cometary asteroids as rapidly disintegrating, short-lived balls of dust loosely held together by a 'glue' of ice; this 'dustball' constitution applies too to many meteor particles of probable cometary origin. As heat percolates into the interior, the body becomes more fragmented until it is unable to hold itself together against any small collision. Development of cracks and fissures is likely throughout the interior as the asteroid weakens; in the presence of slight inhomogeneities such cracks may penetrate 100 metres into the body, and may lead to disintegration of some comets into many fragments of this size at the end of their active lives.

It is likely too that these boulders and cometary asteroids are short-lived. The outgassing observed from the kilometre-sized object Oljato is such that it would last only about 100,000 years before vanishing altogether. A body of similar structure and size 0.1 kilometre, capable of yielding a 50-megaton impact, would last only 5,000 years. There is scarcely time therefore for at least the smaller asteroids to spread away from the Taurid stream before wholesale disintegration is under way; small bodies are therefore likely to be concentrated in the Encke stream. Swarms such as the lunar one may exist for only a few centuries before they disperse along the orbit and it is likely that there exists, within the Taurid stream, an undis-covered asteroid in process of breaking up. It could be up to 30 kilometres across and still have escaped detection. It could also be the prime hazard facing civilization.

In summary, there are many factors which make it likely that devastating impacts are bunched into brief periods of high risk interspersed with longer quiet periods. Tunguska-sized impacts must follow the galactic trend of the large crater-forming ones, so creating periods of high and low risk depending on the galactic environment; within an impact episode the risk is higher if there is a giant comet around, and higher again on those occasions when its orbit intersects that of the Earth; and if large boulders do form in swarms, then during close encounters with the comet or its degassed remnant there is a risk of occasional bombardment on a scale comparable with that of a nuclear war.

Under these circumstances, of course, a re-examination of the mythological and early historical data becomes that much more critical. Even at the conservative baseline impact rate, Tunguskas strike on average every few centuries; 5 or 6 such isolated impacts should have taken place since the time of Christ, with an excellent chance that one of them had an energy of several hundred megatons. The occurence of Tunguska-like impacts in recorded history is therefore expected. In fact, with the temporarily enhanced Tunguska flux implied by the incidence of giant comets and the production of swarms, one can only avoid a sequence of virtually annual impacts (not observed!) by having long periods of quiescence interspersed with brief periods of multiple bombardment. Thus we expect a Dark Age within the last two thousand years. Probably, we had a very near miss at the beginning of the present millennium. The future is equally secure.

COSMIC WINTER

In our scenario we assumed that above a certain energy threshold, so much dust is lofted into the atmosphere that a major climatic catastrophe follows. It takes surprisingly little energy to produce this effect. The application of only 100 megatons or so of energy with 100 per cent efficiency would lift enough fine dust from ground level into the stratosphere to block out sunlight altogether for as long as the dust remained. A large explosion in which some correspondingly smaller fraction of the energy went into lofting dust, or multiple explosions, would therefore have the same effect. This incidentally is a major difficulty with the simple 'dust blocks sunlight' thesis of the 'giant meteorite kills dinosaur' school. Brief stratospheric dusting must have happened many hundreds of times throughout the history of life and cannot therefore be a prime cause of the mass extinction of the dinosaurs. Prolonged dusting from cometary breakup, coupled with worldwide conflagration and blast from cosmic swarms which may include huge asteroids, are another matter. We examine here the relationship between cosmic dust input and climatic disaster.

In 1977, the American astronomer Kowal discovered a faint object which, although it left a short asteroid-like trail on a photographic plate, did not lie in the asteroid belt. Kowal's object (now named Chiron after one of the Centaurs known for his wisdom and goodness) turned out later to be a huge comet in a chaotic, unstable orbit between Saturn and Uranus. Analysis of the orbit shows that the comet passes close to these planets from time to time, having had a close encounter with Saturn in the seventeenth century. During such strong encounters the course of the comet cannot be predicted with accuracy and its orbital history is not well known. Statistical predictions can be made, however, and it turns out that the mean lifetime of Chiron in its present orbit is about 100,000 years, although it could persist for a much longer or much shorter time. When the comet is eventually expelled from this zone, there is a 20 per cent probability that it will be thrown into interstellar space, and an 80 per cent probability that it will be thrown into the region of the inner planets.

Now 100,000 years is a very brief span in relation to the age of the Solar System (4.5 billion years) and even when it was discovered, it was thought most unlikely that we were seeing a satellite newly flung out from the Saturnian system, or an escaping main-belt asteroid. Thus there are only about a dozen asteroids larger than Chiron, which is 250 kilometres in diameter, and the probability that we should be observing one in the act of escaping was very remote. The chaotic orbit, and recent discovery of a coma indicate however that we are dealing with a comet. On the other hand, with Chiron a comet, it is a truly exceptional body, with something like a hundred times the mass of the comet Encke progenitor, and an 80 per cent probability of finding itself in a short-period orbit within 100,000 years. Such an enormous body, disintegrating into dust and asteroids over some thousands of years, would be an awesome sight. But also, it could hardly fail to create a severe climatic trauma – probably an ice age – and a great mass extinction of life would then be unavoidable. It is probably rather fortuitous that Chiron has gone undiscovered for as long as it has, and there are probably not too many Chiron-sized objects waiting in the wings, but a very large population of smaller bodies could well exist, undetected, in the Saturn–Uranus region. Because of our complete ignorance of the size of this population, the hazard is hard to assess and is best judged on the rate at which new comets seem to be entering the short-period population at the present time. A large uncertainty goes with the resulting estimate. But it seems that, on a timescale of 500 years, there is about a 1 per cent risk that another comet, sufficiently large to pose a major threat to civilization, will enter into an Apollo or short-period orbit. The hazard posed by a fresh giant comet entering the Earth's environment is therefore reassuringly small. A much more immediate dust hazard, we shall see, is that posed by the cometary debris already here.

The largest single impact discussed in our story was 1,500 megatons. Even taking an average impact rate say as found from the lunar cratering, thus ignoring the evidence that we are now in a period of enhanced risk, there is a reasonable prospect of such an impact over historical timescales. An impact of this energy would result from a body of mass about 13 million tons coming in at 30 kilometres per second. The atmosphere would ablate an outer skin from the body but would be quite unable to stop the bulk of it hitting the ground. Given the fragile and fluffy composition of cometary material and the energy of the impact, the bolide would immediately disintegrate, any part of it going unvaporized being broken into its constituent interstellar dust particles, which may be about 0.15 microns in diameter: this material would be entrained in the fireball.

In the event of a land impact, the comet fragment would excavate a crater. Experience with surface nuclear bursts[131] indicates that something like a third of a million tons of dust is raised for every megaton of explosion energy. The figure is probably roughly applicable to an impact but may be a few times higher because of the energies involved in an explosion as gigantic as the one we are considering. In the case of a nuclear explosion, a few percent of this dust is made up of sub-micron sized particles, much of it formed by the condensation of vaporized material. To what extent an impact would be comparable with a nuclear explosion is very uncertain (the temperatures reached in an impact are lower but persist longer) and it is not obvious that the nuclear case can be applied unmodified to the impact one. Limited evidence suggests that, out of the ground, something like the mass of the projectile will be excavated in the form of sub-micron particles, and probably several times this amount is produced as vapour condensates. Overall, perhaps 400–2,400 million tons of dust may be generated by the collision, 40–240 million tons of which is in the dangerous sub-micron form. All of the fine material will be entrained in the fireball.

The huge fireball would be too energetic to be contained by the atmosphere. Laden with dust, it would rise upwards at several kilometres per second, incidentally creating powerful winds which themselves sweep up material. Within minutes it would soar to an altitude of several hundred kilometres, in the high stratosphere. Here it stabilizes, and the dust spills outwards.

Closer to the ground, the intense heat of impact would generate fires over a wide area. About 40 per cent of the land surface is forest-covered, and the same again is brushwood, grass and so on. About one-third of the area of the USA is flammable in the summer, for example. A 1,500-megaton impact on a forested region would probably create several hundred fires over an area of 400,000 square kilometres. Depending on cloud cover (clouds might enhance the heat through reflecting the radiation), dryness of the ground and so on, it is possible that these would merge into a single forest fire about 800 kilometres (500 miles) across. Everything combustible along the line of sight of the initial fireball would catch fire and the extent of the conflagration would be limited essentially by the curvature of the horizon. (Prompt loss of life over this area would presumably be considerable.) Adopting figures which have been used to analyse the 'nuclear winter' scenario, one finds that something like 40–50 million tons of smoke would be produced, the smoke particles having diameters about 0.1 micron. Plumes of smoke would rise up to 10 kilometres (7 miles) into the atmosphere.

The situation we have described – some tens of millions of tons of sub-micron dust lofted into the high stratosphere, and a similar mass of smoke pouring into the lower atmosphere, is very similar to that envisaged in the nuclear winter studies, in which the after-effects of smoke and dust from a nuclear exchange, ranging from 100 to 10,000 megatons, are considered. This is *a fortiori* true if the Earth encounters a cosmic swarm. The general similarity is not surprising, given that both situations involve the sudden release of comparable amounts of energy. There are, however, significant differences:

1 Some tens of millions of tons of the comet material is in the form of fragile, sub-micron sized particles ready made for separation and lofting upwards.

2 A single large impact throws a greater proportion of material, higher into the stratosphere, than an equivalent number of smaller explosions.

3 An impact fireball differs in several ways from a nuclear one; it is not clear what effect this has on estimates of dust production.

4 Nuclear war scenarios envisage urban targets with high concentrations of flammable material; a random impact is more likely to strike forest or brush.

It should be emphasized again that great uncertainties attend all these attempts to calculate the dust injection, mainly because of the absence of trials in the field. Arguments have been given for both higher and lower dust inputs in the nuclear winter case. Overriding such uncertainties is the certainty that impacts of vastly greater energy will take place from time to time. From time to time, therefore, Nature will generate its own nuclear winter. Overall it seems likely that during a period of a few thousand years, there is an expectation of an impact, possibly occurring as part of a swarm of material, sufficiently powerful to plunge us into a dark age.

A consideration is that if, as we propose, ancient tales are a valid guide to events in the prehistoric past, then some mention of a climatic glitch should be preserved in the tales of celestial battle. In fact they are, and they are preserved in localities which might be expected to have suffered most from brief, sudden coolings. Myths from the Iranian highlands and the central Asian Altai tell of a golden age which was terminated by a life-destroying cold. In Icelandic myth, for example, one finds the *fimbulvetr*, a terrible winter associated with Ragnarok, the final battle of the gods. In this prophecy of world-end, based apparently on a story lost in time, familiar elements appear: there is the terrible battle of the gods, Odin, Thor and others, accompanied by blazing hosts – the Valkyries, the 'children of darkness, the doom-bringers, offspring of monsters', the sons of Muspell etc. The world ash Yggdrasil, about which the sky rotates, trembles, while:

> Earth sinks in the sea, the sun turns black,
> Cast down from Heaven are the hot stars,
> Fumes reek, into flames burst,
> The sky itself is scorched with fire

and so on. The world ends with snows and terrible winds, or three winters without a summer depending on the source, and there is reference also to an earthquake coupled with the *fimbulvetr*. It may be that such tales do indeed provide a record of a minor, aborted cosmic winter associated with the fireball swarm.

When the dust finally clears, would the climate return to normal? Consider the following sequence of events: suppose, first, that the freezing after an impact is such that large areas of the planet remain snow-covered throughout a summer. With the onset of winter, this snow cover thickens. By the following summer, with dust still suspended in the stratosphere in middle and high latitudes, the snow is too thick to melt; in the following winter it is added to and so on. When, eventually, the dust clears, the land masses of the northern hemisphere are covered in a snowfield which, reflecting sunlight back into space, has become permanent. The snowfield is added to each year until, within a millennium or two, Europe and North America are again covered in ice sheets a kilometre thick, the ocean level has dropped by 50 or 100 metres, and the Earth is locked into a new ice age. Quite a small lowering of temperature, say about 2° or 3° C for some years, would at the present day be enough to cover much of northern Canada, especially Baffin Island (over 1000 miles long), with permanent snow. This in turn would create a cold trough, triggering further cooling and snow over Quebec province to the south (which is largely a high plateau), leading down to the region of New York and the Great Lakes. Once such a feedback got going, it is difficult to see what could stop it.

Whether a new ice age could be triggered by a super-Tunguska impact, acting singly or as part of a swarm, we do not know. Simulation of the mechanism is very complicated because of the many unknowns, and few studies of the problem have been carried out. It turns out that much depends on the amount of sea ice which would form, and where it would drift to. But the movement of sea ice depends on prevailing winds and sub-surface ocean currents, each of which would themselves be strongly affected by the cooling. Penetration of this ice into middle latitudes would extend cold atmospheric troughs southwards, further enhancing the cold in these latitudes. The amount of sunlight reflected back into space from snow-covered forests is another uncertain factor. High latitude cloud cover – another unknown – must also play its part. It is not surprising, given all the uncertainties, that the few mathematical studies which have been carried out on the problem have given widely different answers. To the question, could a super-Tunguska impact trigger plunge the Earth into a new ice age, the only secure answer at present is that we do not know.

We anticipate nevertheless that over and above ice ages, brief episodes of glacial severity will occur at intervals of a few thousand years throughout the few million years of a bombardment episode. These episodes will be marked by abruptness of onset (years rather than millennia). Recovery may well be slowed down by feedback effects. Possibly, even, the Earth will not recover at all, and will be suddenly plunged into an ice age. We should look for these cosmic winters, brief but traumatic coolings between the great glaciations, in the geological record.

They are, in fact, there. In 1972 Dansgaard and his colleagues[132] pointed out that the oxygen isotope 'thermometer' of the last 125,000 years reveals a number of brief, but sudden and drastic, coolings. These are seen in figure 20. These authors, referring to the event of 90,000 years ago, state that 'within 100 yr the climate changed from warmer than today into full glacial severity . . . the drop (in temperature) might have occurred almost instantaneously . . . it took 1000 yr to recover from this catastrophic event'. The temperature drop involved may have been 5–8° C. The same authors also claim that the ending of the last glaciation 10,700 years ago may have come with similar remarkable abruptness, perhaps within 20 years, coinciding with a sharp decline in the concentration of dust in the glacial ice. There may well have been more such events, since anything of less than about 100 years' duration is liable to be missed altogether in the oxygen isotope record. For example the record covering the last 10,000 years looks quite innocuous, but history tells us otherwise.

Fossil pollen provides a record of the prevailing type of vegetation, and hence the climate, in a given locality. In the southern Vosges in France there is a peat bog, the Grande Pile, which yields undisturbed layers of sedimentation laid down, 5 years to a millimetre, since before the last ice age. The palaeobotanist Woillard[133]

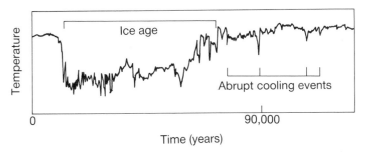

Time (years)

Figure 20. An early temperature record of the last major glaciation, after Dansgaard and his colleagues, from a study of cores extracted from the Greenland ice cap. According to the theory described here, this is essentially a record of stratospheric dusting due to material from the progressively disintegrating Comet Encke progenitor. Note the rapid flickering, the suddenness of onset of individual 'flickers', each a cosmic winter, and the rapidity with which the ice age ended. These sudden changes are not consistent with slow orbital and polar variations as the prime movers, but are expected on the giant comet picture.

found that every interglacial period which she studied in detail ended abruptly; that is, each glacial episode was sudden in onset. The earliest of these 'cold snaps' is shown in detail in figure 21. Over a period of about 150 years there is a steady cooling and then, suddenly, the temperate forest of the region – oak, alder and fir trees – is replaced by pine, birch and spruce, sub-Arctic forest found nowadays in the north of Scandinavia. This happened within twenty years. However because the growth time of trees is of this order, the actual transition of climate may have taken place much more rapidly and could, in fact, have been nearly instantaneous. The mean temperature drop over the northern hemisphere was probably about 5° C. Similar results have been obtained by others, looking at other slices of time.

The geological evidence then, seems to agree reasonably well with the astronomical expectations: at intervals of a few thousand years

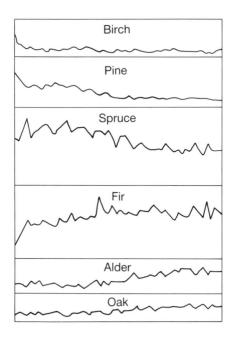

Figure 21. Fossil pollen record in a European peat bog, after G. Woillard. Time moves from right to left and a 300-year period is covered. The extreme left of the record shows a sudden but lasting change in forest composition, from predominantly fir to a colder birch and pine. This sudden cooling of the European climate might have been caused by a stratospheric dusting, from impact or meteoric dust, causing a climatic instability and a change in ocean circulation patterns.

(sometimes more and sometimes less) the planet is abruptly plunged into a glacial climate which may last for a few, a hundred or a thousand years, or which may even lock the Earth into an ice age, from which only slow orbital changes would free it. What would happen if there were a brief blockage of sunlight, lasting even for the few months or years it would take micron-sized dust particles to settle out of the stratosphere? This question has been asked in a nuclear war context, but the results of the climatic and biological studies are probably roughly applicable to the present situation. An important difference is that, whereas in the nuclear-war situation smoke might be scavenged out from the lower atmosphere in a few weeks, in the cosmic case the stratospheric dust would linger. Based on these studies, it seems that a significant cosmic encounter, enough to reduce sunlight to a few per cent of its normal value, would precipitate a rapid plunge to subfreezing temperatures (say $-30°$ C or colder) in continental regions. Freshwater rivers and lakes would freeze over. With the extreme cold would come storms of unprecedented intensity, especially near continental edges. It would be difficult to see even by day, and photosynthesis would come to a halt. The dust raised by one or more large impacts, even if initially confined to one hemisphere, would within a few months spread to the other, and so the effects would be global. The effect on ecosystems would depend on the time of year. An encounter in the summer, corresponding both to the growing season and to passage through the Beta Taurids, would wipe out commercial agriculture for the rest of that year. Thus a drop of only $2°$ to $5°$ C in the northern hemisphere – no more than one observes in the pollen record – would substantially cut wheat and corn production, and might wipe out rice production in Asia. In this situation much of the human race would be wiped out too, and one would not expect civilization to recover for, probably, centuries. A full-blown ice age would of course end civilization in any form we now recognize.

There is a well-known theory, associated with the Yugoslavian Earth scientist M. Milankovitch, according to which ice ages are caused by slow cyclic changes in the Earth's orbit. These changes cause the amount of solar energy reaching the land masses of the northern hemisphere to vary. It is likely that orbital evolution does indeed affect the strength of a glaciation, but unlikely that it is the prime cause. The cyclic variations have been going on continuously throughout Earth history but the ice ages are intermittent. Some additional factor must have triggered the onset of the last ice age, for example. And of course the sudden coolings, usually only years in onset, cannot be explained by orbital cycles which are measured in tens of thousands of years. Looking at figure 20, one may ask why the planet recovered from some abrupt coolings, but not from others. It is likely that, during a comet shower, impacts are continually dusting the Earth, in combination with comet dust. But it is only when the orbital conditions are right that the runaway instability sets in, and the Earth ices over. Calculations show that these orbital conditions exist now.

COSMIC DUST

There is a crucial test for the ideas developed here.[134] If Encke's Comet in its giant phase truly flooded the stratosphere with dust, then that dust should still be identifiable in deposits on Earth. An ideal place to look for the comet dust would be deep within the polar ice caps. Snow falling in these regions becomes compacted into ice, which is laid down typically at a few centimetres a year, and once deposited, is likely to remain undisturbed for the lifetime of the ice cap. Dust, trickling down from above, will become entrained in the ice. We should look for comet dust, laid down in quantities large enough to block out the Sun, and identifiable with the material of Encke's Comet.

Now ice cores, some of them over a kilometre deep, have been extracted from several locations – Byrd Station at the south pole, Camp Century in the north of Greenland, and so on. It has been known for some years that there is an excess concentration of very small particles, many of them sub-micron sized, laid down during the last ice age (figure 22). Over the period around 20,000 to 10,000 BC, during the most recent ice age, the dust concentration seems to be at least ten times higher than the post-ice-age levels. In fact, so much dust was laid down over this period that, if indeed it entered through the high stratosphere, the sky would have been densely veiled, polar sunlight being reduced to a fraction of its present-day level. Of course the effect of a permanent dust veil in the atmosphere is more extreme in polar regions, where the Sun is never very high above the horizon and sunlight has a long path to traverse

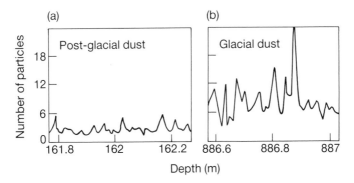

Figure 22. Typical depositions of sub-micron dust particles in polar regions during glacial and post-glacial times (due to Thompson and Mosley-Thompson, 'Microparticle concentration variations'). Note the enhanced dust levels during the last ice age. A significant proportion of this material seems to be cometary in origin.

through the atmosphere. The dust concentration is about that expected on the comet disintegration hypothesis.

There is, however, a hidden assumption in this line of argument. It could be that snow was precipitated at a different rate during the last ice age. If the annual snowfall at the poles was only a tenth of its present rate, then a steady background trickle of dust would become concentrated in the ice without, however, ever having been concentrated in the atmosphere. Fortunately, there are annual variations in the dust concentration, and these can be counted, much like counting tree rings, to give an absolute chronology. These counts show that the timescale shown in the figure is about right. Evidence covering the last 20,000 years, then, seems to show that the polar caps at least were unusually dusty places. Some proportion of this material must be volcanic in origin and some must be wind-blown continental dust (it is likely that the polar regions were windier places than they are now). However the deposition rates at opposite ends of the Earth are surprisingly close, although the distribution of land masses in the two hemispheres is very different.

In an investigation carried out in 1983, the American LaViolette analysed samples of ice from the Camp Century ice core. The technique used, neutron-activation analysis, allows certain elements to be detected in parts per billion. High iridium levels were found in all the samples examined: we are dealing, probably, with cosmic dust. The levels corresponded to deposition rates 10 to 60 times greater than those of the present day. LaViolette found also that the dust had a very unusual chemical composition; and at particular levels in the ice, the overabundances of several elements jumped by large amounts. At one level the concentration of tin jumped by over 27,000 times and silver by over 200 times. It is very unlikely that this is ordinary meteoritic material since there appeared to be particles hundreds of times richer in tin, antimony, silver and gold than ordinary meteorites. It is very unlikely too that this material came from Earth: for example the tin is over 100,000 times more concentrated than the ordinary material of the Earth's crust. LaViolette pointed out that 'If this material is of extraterrestrial origin, its source must be different from that of the majority of meteoritic material.'

If this is material from Encke's Comet, and if a fragment of that comet struck the Central Siberian Plateau in 1908, then any surviving cosmic material from the impact site ought to have the same strange chemical composition as the stuff laid down in the ice caps between ten and twenty thousand years ago. No visible debris was left after that collision. However, under the pressure of the fireball and the subsequent fallout, Tunguska material was impregnated into the peat moss of the area. The Russian Golenetskii and his colleagues have recovered such material and investigated its chemical composition. Their technique was to extract peat cores from the central regions of the impact site, dry them and burn them in a furnace in contamination-free conditions. The final step was a theoretical one, in which an attempt was made to allow for the preferential escape of more volatile elements under the intense heat of the fireball: this last step is the most uncertain, but should give results at least roughly correct. The pre-entry composition of the Tunguska body, it turns out, shows many peculiarities according to Soviet investigators: 'the content of many elements in it differs from their average abundances in the earth's crust, meteorites, and the solar system as a whole'. In particular, there were great overabundances of tin, silver and antimony.

Putting these two studies together, we find the results shown in figure 23. There are caveats. Only trace amounts of elements are being measured, and even with the greatest care contamination of samples is possible: the problem has plagued polar research for the past twenty years. Further, the pre-fireball composition of the

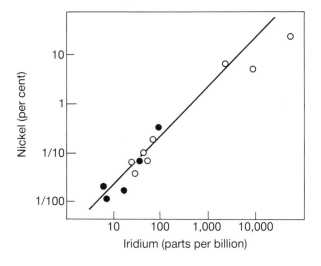

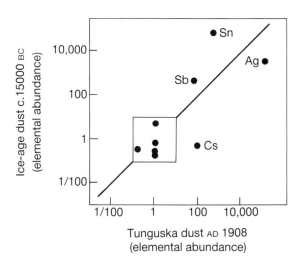

Figure 23. (a) Nickel–iridium ratio for small dust particles laid down on Greenland over the period 19700–14200 BP (filled circles), and from microparticles found at the site of the Tunguska explosion of AD 1908. Both these elements are found in cosmic particles. The good fit of both sources on to the line may imply a common progenitor body, as anticipated from the theory. (b) Comparison of abundances of various elements from the Greenland ice particles and from the Tunguska particles. Material in the box has the 'universal composition' found in unprocessed astronomical bodies. Both dust deposits have identical anomalies relative to the mix – large overabundances of tin (Sn), silver (Ag), Caesium (Cs) and antimony (Sb) – again consistent with a common source, and hence a cometary origin of the Greenland dust.

Tunguska object had to be inferred from the surviving remnants. However it would be very strange if all these effects had conspired to yield identical fictitious anomalies, and we are therefore inclined, pending further measures, to accept the results. The conclusion is remarkable: within the errors, the dust extracted from 10,000-year-old ice a kilometre down in the Greenland ice cap, and that extracted from the moss of the 1908 Siberian impact, have the same exotic chemistry. One should keep in mind too that each analysis was carried out without knowledge of the other, and without knowledge of the astronomical picture we have described. The identity of result supports the hypothesis that the last ice age was caused by dust from a large comet whose debris is still in orbit. We conclude that an encounter with the denser parts of this debris is possible and could have consequences along the lines discussed in our story.

SUMMARY OF RISK ASSESSMENT

There are three main reasons why the magnitude of the cosmic impact hazard has not been appreciated until now. First, the average impact rate in the 10–10,000 megaton range has in the past been underestimated by a factor of about ten. Second, only the local effects arising from blast have been contemplated. Third, there is the new consideration that the impacts may be strongly bunched in time, on various scales from the geological downwards, so leading to periods of high risk interspersing low risk. Arguments have been given to indicate that we are currently in a higher-than-average risk period. Our 'best estimates' of the cosmic hazard are then as follows:

1   The environment of the Earth is substantially more hazardous than has been realized until now. Several risks – nuclear error, urban impact, stratospheric dusting – have been identified.
2   These risks are measured at an average level of a few per cent within a human lifetime, but in reality they may be concentrated into brief catastrophic periods in which multiple impacts occur.
3   Especially given the sensitivity of modern agricultural systems to temperature and the arrival of the Beta Taurids in the northern hemisphere growing season, the prospect of global catastrophe from stratospheric dusting appears to be the greatest natural hazard.[135]

4 These hazards are substantial when considered on the timescale of human civilization: should they materialize, civilization might be plunged into a new Dark Age, and it is even possible that the human race would come to an end.

Epilogue

It is the fate of all species to become extinct and most manlike species have already done so. Over and above extinction, large population fluctuations take place in nature, sometimes within a few years. The controlling factor is often climate, and Earth's climate, in turn, can be greatly affected by its astronomical surroundings.

The two and a half centuries which lay between the Gervase chronicle of 1178 and the onset of the Black Death in Europe in 1348 saw 'an acute crisis developing in human affairs'. One chronicler[136] at least reports of the most immediate cause of the plague in 1345 that 'between Cathay and Persia there rained a vast rain of fire; falling in flakes like snow and burning up mountains and plains and other lands, with men and women; and then arose vast masses of smoke; and whosoever beheld this died within the space of half a day . . . .' There seems little doubt also that a worldwide cooling of the Earth played a fundamental part in the process. The Arctic polar cap extended, changing the cyclonic pattern and leading to a series of disastrous harvests. These in turn led to widespread famine, death and social disruption. In England and Scotland there is a pattern of abandoned villages and farms, soaring wheat prices and falling populations. In Eastern Europe there was a series of winters of unparalleled severity and depth of snow. The chronicles of monasteries in Poland and Russia tell of cannibalism, common graves overfilled with corpses, and migrations to the west. Even before the Black Death came, then, a human catastrophe of great proportions was under way in late medieval times. Indeed the cold snap lasted well beyond the period of the bubonic plague. A number of such fluctuations are to be found in the historical record, and there is good evidence that these climatic stresses are connected not only with famine but also with times of great social unrest, wars, revolution and mass migrations.

In spite of their traumatic effects, these global coolings probably amounted to no more than about a degree in average summer temperatures as compared with today: even relatively minor climatic effects have had a profound influence on human history. A major cosmic winter, on the other hand, is likely to produce a rapid global climatic cooling amounting to several degrees. With the modern dependence on 'green revolution' crops, finely tuned to give a high yield under a narrow range of climatic conditions, the onset of such a 'winter' would cause the population of the world to crash in the course of a decade, or even a single year. Such events are completely outside normal experience and their existence is not generally recognized, even though they represent a hazard vastly more horrific than any of the more familiar catastrophes such as earthquake, famine or flood. As we have seen, there is a fair chance that our planet will be caught in the icy grip of such a cosmic winter at intervals of a thousand years or so. Lesser catastrophes, little ice ages in effect, will arise on timescales of a few centuries. At an individual level, the risk amounts to a few per cent within a lifetime. More to the point, though, civilization is in the presence of a hitherto unrecognized cosmic phenomenon which could plunge it without warning into a Dark Age.

What can be done? Unfortunately the extent and epoch of the next cosmic winter depend for the moment on a number of imponderables which lie outside the scope of existing knowledge: it is not now possible to make an accurate assessment of what the future has in store. This is clearly not a satisfactory state of affairs, nor can we expect that Nature will hold back on account of our ignorance or lack of preparedness. However, in view of the seriousness of cosmic winters for human survival, and noting the vast expenditures to the tune of many billions of dollars on a whole variety of preparations for all manner of lesser hazards and calamities, both man-made and natural, disease and nuclear war not excluded, one must surely note also that not a single cent (or penny!) of taxpayers' money is currently devoted to their study.

The first step must therefore be one of exploration. An asteroid in a Taurid orbit, carrying 100,000 megatons of impact energy, coming out of the night sky, would be visible in binoculars for about six hours before impact. By the time it was a naked-eye object it would be at most half an hour from collision. In its final plunge it would be seen as a brilliant moving object for perhaps 30 seconds. One needs more time than this to prepare for the winter. A thorough exploration of the Earth's surroundings, and the discovery and tracking of probably tens of thousands of bodies, is therefore a first requirement. This is technically feasible.

Complementing such an observational programme, a fresh exploration of the past, armed now with the new astronomical understandings, is also necessary; not just for its own sake but also to arrive at a better understanding of the risks. For example one would like to find physical evidence to confirm or deny an astronomical catastrophe in the Near East in the second millennium BC. A search for Tunguska-like material in the Mediterranean sea bed is one possibility. Evidence for high-temperature effects is another: the brief thermal pulse from a big fireball would glaze rocks and create shadow effects, which fire created by straw-burning invaders would not. A thin layer of soot, arising from extensive conflagration on open ground, should be present in strata from the period. The effects of blast might be distinguished from those of earthquake by the disposition of fallen debris; and so on. To go from mere statistical projection to detailed forecasting, then, a generation of exploration, both of the Earth's environment and of our history and prehistory, will be necessary. As we have remarked, such studies cannot be seen only as an academic game: there is nothing academic about a 1,000-megaton impact, and the modern prospects for nuclear error, not to mention nuclear meltdown, exacerbate the issue.

And if the sirens should sound, what then? It may be marginally within the capacity of present-day technology to divert a small asteroid, given enough warning, though not a swarm of them: it would take a heavy launcher and a hydrogen bomb. But at least, unlike our forebears, we have a chance to act; we need no longer be helpless in the hands of the gods. The main problem at the moment is to be aware that there is a problem.

Three thousand years ago, in accordance with age-old practice, the kings of Babylon were still employing astronomer-priests to give warnings of cosmic visitations. A thousand years ago, the emperors of China were still relying on similar skills, while in Europe the Pope saw messages in the sky and urged Holy War. But this latter was an aberration; for the last two and a half thousand years have seen the decline and fall of the sky gods, and the growing presumption that the cosmos is stable and regular. The shift of paradigm has been unconscious, convenient, insidious and thorough. Probably, the rediscovery of a lost tradition of celestial catastrophe could not have been made through analysis of ancient texts alone; a key had to be provided, and it has been, by the paraphernalia of modern science. It is a salutary lesson both on the capacity of human reasoning to get it wrong for long periods of time, and on the essential unity of knowledge.

It would be naïve to think, however, that one merely has to point to deep-seated cracks in the structure of modern knowledge, to have scholars setting to and constructing a better framework within which mankind might plan his future. There is considerable intellectual capital invested in the status quo, enough to ensure that those with an interest in preserving it, the 'enlightened' and the 'established', will continue to present the cosmos to us in a suitably non-violent form. The history of ideas reveals that some will even go further and act as a kind of thought police, whipping potential deviants into line. For them, temporal power takes precedence over the fate of the species.

There is a need for this book.

Notes

1   Thomas, *Religion and the Decline of Magic*.
2   Cohn, *The Pursuit of the Millennium*.
3   Homer, *The Iliad*.
4   Virgil, *The Aeneid*.
5   Lucretius, *De Rerum Natura*, translated by R. E. Latham (Penguin Classics, 1951), copyright © R. E. Latham, 1951. The italics are ours. They emphasize specific aspects of the real world which appear to have been clearly recognized in classical times and which are now confirmed by modern science, as we shall see.
6   *Sybilline Oracles*.
7   Seneca, *Naturales Quaestiones*. The italics are ours.
8   Tian-shan, 'Ancient Chinese Observations' entry no. 39. The implied date is 22 May 12 BC, the time 3–5 pm. A *fou* is an earthenware pot; a *zhang* is 12 degrees.
9   Savage, *The Anglo-Saxon Chronicles*.
10   Morris, *Historical Tales: the Romance of Reality*. This impressive extract is repeated in the posthumous publication by Velikovsky (1982), wherein it is held that mankind is *subconsciously* traumatized by past human experience into minimizing or even trivializing celestial dangers. The alternative explored in this book, that celestial danger was programmed out of the body of *conscious* knowledge for spurious intellectual reasons, has much the same effect, however.
11   Jacobsen, 'Mesopotamia'.
12   Frankfort, H. and Frankfort, H. A., Myth and Reality.
13   Davidson, *The Stars and the Mind*.
14   Butterfield, *The Origins of History*. This study was also published posthumously but, being incomplete, without the benefit of the author's considered conclusion. Nevertheless, the insight emerging from this historical survey, conducted without any reference to the recent astronomical findings, suggests a broad concurrence of outlooks will not be difficult to achieve.
15   Neugebauer, 'The history of ancient astronomy'.
16   Van der Waerden, *Science Awakening II*.
17   This understanding of the Universe was perfected during the subsequent half-millennium, achieving its most purified expression in Ptolemy, *The*

Almagest. Its purpose cannot be understood however without reference to Ptolemy, *The Tetrabiblos*.

18 Whilst it is possible to argue that the interbreeding of civilizations, arising for example as a consequence of mass migration, may be the most significant process underlying the course of history (cf. Darlington, *The Evolution of Man*), we emphasize here that particular extreme environmental effects are likely to be more fundamental.

19 Oates, *Babylon*.

20 See for example Needham, *The Grand Titration*.

21 The Chinese astrological record, like its Babylonian predecessor was put together over a period of nearly two thousand years and was eventually made known in the west following its translation by the French physicist, Biot, 'Catalogue générale des étoiles filantes'. The record of observed 'guest stars' has recently been revised and updated by Ho, 'Ancient and medieval observations', but no comparable study of ancient meteor observations yet exists. A considerable understanding of the spirit of astrology in China may be gained from Schafer, *Pacing the Void*.

22 Kuhn, *The Structure of Scientific Revolutions*. The author argues not only that knowledge advances in much the way we have described, by way of sudden paradigm switches, but also that the choice of fundamental hypotheses at any one time tends to be socially motivated and sustained.

23 A landmark in the study of historical climate is provided in the comprehensive survey by Lamb, *Climate: Present, Past and Future*. A detailed study of the evidence relating to recent climatic recessions is to be found in Grove, *The Little Ice Age*.

24 See for example Simmons and Tooley, *The Environment of British Prehistory*. For a broader perspective, see Goudie, *Environmental Change*. The more or less steadily ameliorating post-glacial climate suddenly went into decline around 3500–3000 BC with a sequence of subsequent recoveries and reversals of varying intensity. This very marked decline was characterized in Britain for example by a sudden loss of tree cover, changes in vegetation, widespread evidence of burning revealed by microscopic carbon particles, and subsequent peat bog formation, a pattern that recurs to a significant degree in the century or so around 1200 BC. The idea that these clearances were the work of man has been embraced by experts unaware of potential natural agencies capable of inducing fire.

25 Such difficulties of interpretation are now being rapidly solved by the use of measurements of certain stable and short-lived isotopes (e.g. ^{2}H, ^{10}Be, ^{14}C, ^{18}O) in minute samples of 'historical' materials. Fluctuations in the concentrations of these isotopes evidently correlate with global climate. Besides revealing major ice ages every 100,000 years or so, these measurements also indicate serious climatic recessions lasting a century which may be no more than a millennium apart. The fluctuations during the last 10,000 years show, in addition, a variety of underlying shorter-period

cycles which seem to be reflected in the Sun's behaviour as well. Unfortunately there are several possible physical processes capable of producing the fluctuations and it remains to be established whether the Sun, a terrestrial reservoir such as the ocean, or a periodic dust source in interplanetary space, is the primary agent. Nevertheless these cycles and their explanation are evidently a major key to the long-standing mystery of the climate and weather, and can be expected to bear their final fruit in the not-too-distant future. The proceedings of a Royal Society Discussion Meeting in London during February 1989 provide an up-to-date account of the present state of research in this field.

26 Renfrew, *Before Civilization*. For a contrary view by an archaeologist, attributing to astronomical and environmental factors, rather than sociological ones, the dominant roles in prehistory, see McKie, *The Megalith Builders*. A well-argued commentary on the 'sociological' approach of many archaeologists is to be found in Bradley, *The Social Foundations of Prehistoric Britain*, which concludes with the remark 'It is time that archaeologists accepted that they can recognize patterns which they had not expected to see'.

27 Wilson, 'Egypt'. For a definitive study of sacral kingship, see Frankfort, *Kingship and the Gods*. Some useful supplementary texts on ancient Egypt are: Gardiner, *Egypt of the Pharaohs*, David, *Cult of the Sun* and Rundle-Clark, *Myth and Symbol*.

28 See, for example, Fraser, *The Golden Bough*, and *The New Larousse Encyclo-pedia of Mythology*. Introducing the latter book, Robert Graves writes that 'mythology is the study of whatever religious or heroic legends are so foreign to a student's experience that he cannot believe them to be true. Hence the English adjective "mythical" meaning "incredible"; and hence the omission from standard European mythologies, such as this, of the Biblical narratives even when closely paralleled by myths from Persia, Babylonia, Egypt and Greece'. We take the view that most cosmic elements of Biblical narrative and mythology (often referred to as creation and combat myths) should be understood as deriving from real though not necessarily comprehended astronomical phenomena. That this might be so was recognized earlier this century in books by Bellamy, *Moons, Myths and Man*, and Velikovsky, *Worlds in Collision*, though neither author was able to provide a valid scientific rationale. Excellent introductions to creation and combat myths are found in: Blacker and Loewe, *Ancient Cosmologies* and Forsyth, *The Old Enemy*.

29 Plutarch, *Concerning Isis and Osiris*. Typhon was explicitly recognized as a comet in late antiquity; of the nine types of comet described by astrologers such as Lydus (third century AD), one was 'typhonic'.

30 The determination of absolute chronology in the pre-Christian era depends on a wide range of studies which we do not attempt to survey here. Suffice to say that most Near Eastern chronology BC is currently established through precise cultural and historical synchronizations with

a year-count based on Egyptian king-lists and the so-called Sothic calendar. Most Central European chronology BC, on the other hand, is established through intrinsically less precise carbon-14 synchronizations with a year-count based on dendrochronological sequences. These independent scales are apparently in broad agreement to a few per cent over 5,000 years (Mellart, *Egyptian and Near Eastern Chronology*), but the possibility of a larger deviation over one or more extended periods cannot be excluded, especially if ambiguities are still present in the Egyptian king-lists (*cf.* Gardiner, *Egypt of the Pharaohs*, Appendix). Severe climatic depressions on a global scale (indicated by concentrations of extremely narrow tree rings within periods of less than 20 years and by simultaneous acidity peaks due to atmospheric dust veils) seem to have occurred at epochs: 4375, 3195, 1626, 1150 BC and AD 540 ± 50 years, without a known volcanic association in every case (*cf.* Baillie and Munro, 'Irish tree rings',); the anti-correlation with known periods of high civilization during proto-history is evidence for a significant role in history on the part of climate.

31 Kitto, *The Greeks*. For a lively archaeological snapshot, see Wood, *In Search of the Trojan War*.

32 Waddell, *Manetho*.

33 Recent research may be indicating that the epoch of the climatic depression *c.* 1625 BC (note 30) coincides with the demise of the Minoan civilization and is associated with the second rather than the first Cretan palaces, i.e. all Minoan dates before, say, 1400 BC need to be set earlier by approximately 75–125 years.

34 Sandars, *The Sea Peoples*.

35 Graves, *Greek Myths*. For a review of Hesiod's sources in particular, see Walcot, *Hesiod and the Near East*.

36 Herodotus, *The Histories*.

37 Guthrie, *Orpheus and Greek Religion*. The cult's astronomical associations have been discussed by Van der Waerden, *Science Awakening II*.

38 Ovid, *Metamorphoses*.

39 Dall'Olmo, 'Latin terminology relating to aurorae, comets, meteors and novae'.

40 The current orthodoxy is that Anatolian, Babylonian and Iranian influences were the principal ones affecting the very earliest Greek theogonies: see for example Walcot, *Hesiod and the Near East* and West, *Early Greek Philosophy*. It is not intended to suggest any Egyptian influence, for example through Orphism, beyond that which has been established through textual analysis, but merely to emphasize the detailed features that the respective cosmogonies have in common. See also James, *The Ancient Gods*.

41 Neugebauer indicates that there is no evidence for any serious Egyptian influence on Greek *mathematical* astronomy. See his 'The history of ancient astronomy'. However the Egyptians were skilled in practical

geometry, as evidenced in previous generations by their construction of
pyramids and temples, and possessed a reasonably matter-of-fact natural
philosophy, as evidenced by Plato's testimony (chapter 5). Thus the
apparent lack of influence is not necessarily an adverse comment on
Egyptian scientific thought and practice since it is now known that the
Greeks forced planetary astronomy into a false geometrical mould of
deferents and epicycles.

42 This quotation is taken from a hymn to the sun-god Aton, possibly written
for the pharaoh (Akhenaton) himself. See Pritchard, *The Ancient Near East*.

43 This summary closely follows James, *The Ancient Gods*.

44 The account here is based on Plato, *Timaeus and Critias*, translated by
H. D. P. Lee (Penguin Classics, Revised Edition, 1971) copyright ©
H. D. P. Lee, 1966, 1971. That there should come to us on the authority of
one of the greatest philosophers of all time the knowledge that a vast
island continent was once lost in a massive catastrophe has never ceased
to amaze; but that there should be, after centuries of study, still no idea at
all when or where the great event took place certainly puts the mystery in a
class of its own. The suggestion that Plato's Atlantis is *celestial* has to our
knowledge been made only once before in modern times, by Prof.
H. A. T. Reiche of the Massachusetts Institute of Technology: see note
98. However the idea was not unknown in classical times, being
mentioned in the writings of the third-century author Amelius. Reiche
suggested that 'the traumatic folk memory of [a] seismic catastrophe of
1200 BC may well have coloured Plato's description of Atlantis's end –
without the latter catastrophe being reducible to the former', whilst also
linking the seismic event with the great eruption of Thera on the Mediterranean island of Santorini. However as we shall see, the same basic tale of
catastrophe recurs from Scandinavia to China, far from the influence of
this eruption. Separate celestial events in 1626 and 1150 BC (note 30),
observed worldwide, are now more probable.

45 This sentence is the first known admission of physical 'action at a
distance' without an apparent intermediary. The reference here to the
role of shepherds is unlikely to be entirely casual and may well hint at an
ancient Babylonian source for the knowledge (chapter 1), whilst also
offering a clue as to why there is so much confusion over the date (*cf.*
Folliot, *Atlantis Revisited*). In the Book of Genesis, it may be recalled, the
lengths of the patriarchs' lives are measured in hundreds of years.
Similarly, in the ancient Babylonian records found on clay tablets in
Mesopotamia, it is said that 23 kings had reigned in the Land of Two
Rivers after the Flood and that the total of their reigns exceeded 24,000
years. These are clearly not realistic human lifespans but they are
characteristic of the visible survival time of short-period comets, and
many of the problems of interpretation that arise in seeking to
comprehend these early genealogies are considerably reduced if we
recognize their fundamentally celestial origin. In like fashion Athens'

prehistoric kings were probably celestial deities, Cecrops and Erechtheus being depicted by the Athenians themselves as partly or entirely serpentine.

46 Described in Plato, *Timaeus and Critias*, p. 49. The notion of fiery rings is also to be found in the works of Anaximander (*cf.* West, *Early Greek Philosophy*). In general, however, modern commentators have been in great difficulty trying to turn these bands in the sky into imaginary lines, e.g. see Dicks, *Early Greek Astronomy* and Heath, *Aristarchus of Samos*.

47 How cosmological theory evolved over the period 700 BC to AD 300 is too large a subject to describe in any detail in this book – our aim is merely to highlight certain issues whose significance is commonly overlooked. See also Bailey et al. *The Origin of Comets*. For a somewhat broader perspective, see Butterfield, *The Origins of History*, Cornford, *Principium Sapientiae*, Cronin, *The View from Planet Earth*, Davidson, *The Stars and the Mind*, Farrington, *Greek Science*, and King, *The Background of Astronomy*. The general historical background is well covered in, for example, Boardman et al., *The Oxford History of the Classical World*.

48 Aristotle, *Meteorologica*. See also Jaki, *The Milky Way*.

49 See Cornford, *Principium Sapientiae*.

50 See chapter 1 and Neugebauer, 'The history of ancient astronomy'. Neugebauer appears to have been the first to realize this was a revolutionary rather than an evolutionary development. Paradoxically, the discovery seems to have had the unintended effect in post-war years of casting the pre-Socratic natural philosophers in a non-scientific mould, at least so far as many scientists are concerned.

51 A much debated issue now but the tendency during the later classical period to 'save appearances' rather than establish truth is well recognized: e.g. see Farrington, *Greek Science*. Newton, *The Crime of Claudius Ptolemy* casts blame at a particular individual when it could perhaps be more widely applied. See also Hetherington and Ronan, 'Ptolemy's Almagest'.

52 Whiteside, 'Before the Principia'. Newton apparently began his investigation of planetary motions very much in the general spirit of Descartes' vortex theory, replacing the familiar crystalline spheres driving the planets with a transparent fluid which, in its subsequent nineteenth-century manifestation, became the conceptual basis of the physical aether. In the event, the properties of this fluid proved something of a mystery and Newton was forced against the spirit of cartesian thought to describe planetary orbits in terms of an 'action-at-a-distance' gravitational law. Indeed the success of this law was an embarrassment on several accounts for, on applying it to comets, the latter's catastrophic potential was evident, causing Newton to become a secret catastrophist: hence Keynes's dictum, often thought somewhat surprising, that Newton was the last of the Babylonian astrologers (Keynes, *Newton, the Man*).

53 See Farrington, *Greek Science*; also Smart, *The Religious Experience of Mankind*.

54 Bailey et al., *The Origin of Comets*; Cornford, *Principium Sapientiae*.

55 Smart, *The Religious Experience of Mankind*.

56 In our discussion of the Exodus story in *The Cosmic Serpent*, we associated these cosmic forces with the progenitor of the Taurid meteor stream and Comet Encke. That the pillar of fire was a comet has been suggested from time to time at least as far back as the Middle Ages, and we were able to demonstrate that the appearance and behaviour of the pillar of fire are indeed those expected during a close encounter with a great comet. More recent research has however raised the possibility that Comet Halley was the culprit. Securely established records of Halley's comet go back only as far as 235 BC, although it must have been one of the most spectacular phenomena of antiquity during its 76-yearly returns. As it happens, Comet Halley had a remarkably close encounter with the Earth in September 1404 BC (Yeomans and Kiang, 'The long-term motion of Halley's Comet'). So close was the passage to the Earth that the gravitational deflection could not be calculated with certainty and it was not possible therefore to track its orbit further back in time. Very probably, the Earth passed deep within the tail of the comet, which must have been a terrifying sight. No closer encounter occurred in the pre-Christian era. Furthermore, within the uncertainties of the calculation, 1404 BC is the time of the Biblical Exodus. It is conceivable therefore that the Exodus story contains the earliest record of Comet Halley.

57 James, *The Ancient Gods*.

58 Baigert et al., *The Holy Blood and the Holy Grail*.

59 Cohn, *The Pursuit of the Millennium*.

60 Augustine, *Concerning the City of God against the Pagans*.

61 The precise circumstances relating to the strange collapse of sub-Roman Britain in the fifth decade of the fifth century AD and the revival of supposed Arthurian Britain in the century following has been reviewed by several authors: see Myres, *The English Settlements*, Morris, *The Age of Arthur*, Fisher, *The Anglo-Saxon Age*, Salway, *Roman Britain* and Alcock, *Arthur's Britain*. Translations of original texts are now readily available, for example: Winterbottom, *Gildas*, and Morris, *Nennius*. Further insights concerning celestial associations of Arthurian history are obtainable from Geoffrey of Monmouth's *History of the Kings of Britain*, and Tolstoy, *The Quest for Merlin*. Celestial deities have been recognized in Arthurian legend since the nineteenth century, e.g. Rhys, *Celtic Heathendom*. The suggestion that the 'ruin of Britain' was caused by an exceptionally violent fireball explosion remains to be proved. In the London *Penny Magazine* of 1834, there occurs the following strange report: 'in the process of draining the Isle of Axholme in Lincolnshire, evidence has everywhere been found not only of previous vegetation but that this spot must have been suddenly overwhelmed by some violent convulsion of nature. Great numbers of oak, fir and other trees were lying 5 feet underground.' It was reported that the tree trunks were all aligned north-west/south-east and had not been

'dissevered by the axe but had been burnt asunder near the ground, the ends still presenting a charred surface.' The resemblance to the scenes of devastation at the site of the 1908 Tunguska impact (chapter 11) is striking.

The Isle of Axholme is a low-lying area a few square miles in extent, east of Doncaster. It is part of the Midland/East Anglia region already mentioned where the Roman civilization appears to have been erased. If there was a cosmic disaster in the fifth century AD in this region, it should now be discoverable by direct archaeological investigation.

62   Klinkerfues, *Gottinger Nachtrichten*, tr. by Fisher, *Popular Astronomy*. Two sources are quoted: 'In this year (524 after the birth of Christ), though, there occurred also much running of the stars from evening quite to daybreak, so that everybody was frightened, and we know of no such event beside'; and 'For twenty days there appeared a comet, and after some time there occurred a running of the stars from evening till early [morning], so that people said all the stars were falling'. The year AD 585 saw this shower recur; it is recorded again in 837 (China) and 899 (Egypt), and is thereafter lost until 1584. As described in chapter 9, Biela's comet, whose orbit passed within a few Earth radii of the Earth's, was discovered in 1826, split in two in 1846, and had vanished by 1865.

63   The point to be made here is that there has long been a disposition on the part of historians to take no notice of the medieval concern for gods, comets and catastrophes. Dealing with the events on the continent in the aftermath of the Arthurian age, Gibbon in *The Decline and Fall of the Roman Empire*, for example, reports the greatly heightened concern at the appearance of several bright comets, followed by earthquake and fire, during the reign of Justinian: 'the nations, who gazed with astonishment, expected wars and calamities from their baleful influence; and these expectations were abundantly fulfilled'. But the absurdity of any connection was already presumed and Gibbon goes on to say, rather aimlessly, that 'the astronomers dissembled their ignorance of the nature of these blazing stars, which they affected to represent as the floating meteors of the air; and few among them embraced the simple notion of Seneca and the Chaldeans that they are only planets of a longer period and more eccentric motion'.

64   Bronsted, *The Vikings*.

65   Riley-Smith, *The First Crusade*.

66   The Domesday Book was so named, it seems, because British subjects were convinced their Norman conquerors were preparing for world-end. Following the decline of the eleventh-century fireball flux, there followed two centuries of increasing prosperity and growth during which the Christian church generally enjoyed the confidence of the European population (see Southern, *Western Society and the Church*). These equable conditions eventually went into sudden decline in the circumstances rather luridly described by Morris, *Historical Tales*.

67   See for example King, *The Background of Astronomy*, and Cronin, *The View from Planet Earth*, for more extended historical accounts.
68   Hellman, *The Comet of 1577*.
69   Stecchini, *The Newton Affair*; see also Clube and Napier, 'Mankind's future, an astronomical view'.
70   The 1833 Leonid shower, taking place in the era of newspapers, yields a more complete picture of the effect of a meteor storm on the population below: see Sanderson, 'The night it rained fire'. Over 200,000 meteors fell over a nine-hour period. According to Sanderson, 'the most terrifying aspect of the shower was the many brilliant fireballs'. Some of these were as bright as the full moon, and many people were wakened by the flashes of light thrown into their bedrooms. Ten to fifteen smoke trails from the bright fireballs were often visible at a time. Most of the meteors were faint, fast-moving, apparently coming in waves, and were too numerous to count: 'never did rain fall much thicker than the meteors fell towards the earth'. There was a widespread belief that 'the world is now actually coming to an end, for the stars are falling'.
71   With Victorian confidence in Newtonian physics went the solid under-standing that space is filled with an invisible and frictionless material substance which mediates in transmitting the forces which physicists now attribute to the action of 'fields'. This idea ran into severe difficulties when laboratory experiments designed to measure absolute motion through the supposed medium failed to work. Although Lorentz demonstrated that the null result was understandable if bodies moving through the medium were suitably contracted in the direction of motion, Einstein realized that the implied departures from Newton's laws could also be understood in terms of the so-called theory of relativity. By this time, physicists were so exasperated with their failure to devise a proper working model of the pervasive medium that they were persuaded to adopt the Einsteinian rather than the Lorentzian theory. This theory was fundamentally very different from Newton's but predicted exactly the same equations in circumstances where Newton's laws were known to work, and so physicists managed to preserve the impression of an evolutionary rather than a revolutionary development of twentieth-century from nineteenth-century physics. It is a moot point whether the impression is false.
72   Although the existence of giant comets has been recognized by astronomers for some time, the importance of knowing how they behave when trapped in very short period orbits near the Earth has only been properly appreciated within the past few years, specifically in the authors' first book *The Cosmic Serpent* and their paper, 'The microstructure of terrestrial catastrophism'. The extraordinary implications of this dis-covery, some of which are confronted in this book, are just beginning to be realized. The precise orbit of the Beta Taurid swarm is not at present known; however, see Clube and Asher, 'The Evolution of Proto-Encke'.

73 Hart, 'The evolution of the atmosphere of the earth'. Various feedback effects, especially a systematic change in cloud cover, might act to modify these figures.

74 At the present time it is widely held that comets are formed on the fringes of, or just beyond, the planetary system. That comets might on the contrary be genuinely interstellar in origin has been proposed by a number of workers, including the authors. Comet cosmogony raises many complex issues which have been reviewed by Bailey et al., *The Origin of Comets*.

75 Rudaux and de Vaucouleurs, *Larousse Encyclopedia of Astronomy*; Flammarion and Danjon, *The Flammarion Book of Astronomy*.

76 Kronk, *Meteor Showers*. This book provides an excellent descriptive survey of observed meteor showers and their properties.

77 This is a group of about a dozen known comets, with orbital periods 500–1,000 years, in very eccentric though similar orbits which bring them very close to the surface of the Sun. There could be a hundred or so comets altogether in the group. Some members of this group have been very large objects (e.g. the comets of 1843, 1880, 1882 and 1887). The comet of 1882 broke into 4 or 5 distinct fragments, and the comets appear to arrive bunched in time. It is very likely that these bodies are the remnants of another giant comet which has undergone a series of disintegrations. They do not, however, pass close to the Earth. We have, then, observational evidence of two exceptional comets which were thrown into periodic orbits from the last disturbance of the comet cloud, namely the progenitor of the Sun-grazing group and the progenitor of Encke's Comet and the Taurid streams. However, there are undoubtedly more, at present on the fringes of the planetary system, and with the potential to cause glaciations if thrown into the inner planetary system (chapter 17).

78 Astapovic and Terenteva, 'Fireball radiants of the 1st–15th centuries'.

79 Russell et al., 'Interplanetary magnetic field enhancements'.

80 Stohl, 'On the distribution of sporadic meteor orbits'.

81 Whipple and Hamid, 'On the origin of the Taurid meteor stream'.

82 Kresak, 'Sources of interplanetary dust'.

83 An account of Leonard Kulik's investigations is given by Krinov, *Principles of Meteoritics*.

84 Deacon, 'The 1908 Tunguska explosion'. An impact energy of 10 megatons is often quoted for this event, based on the extent of the flattened forest. However, estimates based on this are somewhat unreliable and we have preferred the barographic calculation, which is more robust.

85 The Tunguska bolide approached from a roughly southern direction on a long, shallow trajectory, approaching from sunwards. Its orbit in space can be reconstructed from sightings of the fireball along its track, and this led the Czechoslovakian astronomer Kresak, 'The Tunguska object', to suggest that the missile was part of the Taurid meteor complex. This was disputed by Sekanina, 'The Tunguska event', though on grounds which

no longer seem to be valid (note 72), and supported by Levin and Bronshten, 'The Tunguska event'.

86   Hartung, 'Was the formation of a 20-km-diameter impact crater on the Moon observed on June 18, 1178?'. References to the Latin source and translation by Richmond Y. Hathorn are given therein.

87   Callame and Mulholland, 'Lunar crater Giordano Bruno'.

88   The suggestion that the Giordano Bruno crater was due to the Taurid swarm which the Earth–Moon system encountered in 1975 was first made to a public audience by Dr K. Brecher at a talk given to the American Astronomical Association in Baltimore, Maryland on 10 June 1984.

89   Useful general texts reviewing recent developments in the study of ancient astronomy, with particular emphasis on archaeological and mythological aspects, are to be found in Aveni, *World Archaeostronomy*, Blacker and Loewe, *Ancient Cosmologies*, Heggie, *Megalithic Science*, and Krupp, *In Search of Ancient Astronomies*, *Echoes of the Ancient Skies* and 'Archaeostronomy and the roots of science'.

90   James, *The Worship of the Sky God*.

91   *Apollodorus*.

92   Van der Waerden, *Science Awakening II*.

93   Fuhr, *Ein Altorientalisches Symbol*.

94   Silk paintings unearthed from Han tomb no. 3 at Mawangtui, China, 168 BC.

95   See Clube and Napier, *The Cosmic Serpent*, and Hadingham, *Circles and Standing Stones*.

96   Jacobsen, *The Treasures of Darkness*.

97   Personal communication from the American anthropologist E. C. Baity. See also Baity, 'Archaeostronomy and ethnoastronomy so far'.

98   Reiche, 'The language of archaic astronomy' and references therein.

99   de Santillana and von Dechend, *Hamlet's Mill*.

100   Pliny, *Natural History II*.

101   Nonnos, *Dionisiaca*.

102   Huxley, *The Way of the Sacred*.

103   Mackenzie, *Indian Myth and Legend*.

104   Whitlock, *In Search of Lost Gods*. See Frazer, *The Golden Bough*, for an interesting account of fire rituals; also Note 97.

105   To the 'slaughter of livestock before winter' we should add 'the onset of the dry season' as another local explanation for this worldwide celebration (Aztec rather than Celtic): see Milbrath, 'Star gods and astronomy of the Aztecs' and Krupp, 'The "binding of the years"'. Certainly by the time of the last New Fire ceremony (AD 1507, near Mexico City) the Taurids were relatively inconspicuous; however the 260-day calendar had been known to the Mayas for 1,500 years, under the name *tzolkin*. The calendar therefore had its roots in, and survived through, periods of intense Taurid fireball activity.

106   Holmes, *The Age of the Earth*.

107 For references, see Clube and Napier, 'Giant comets and the Galaxy'. An authoritative review of the whole question of terrestrial periodicities as understood up to about 1980 is given in McCrea, 'Long time-scale fluctuations'. The recent bout of rediscovery of terrestrial periodicities amongst Earth scientists begins with Raup and Sepkoski, 'Periodicity of extinctions'. See also Sepkoski, 'Periodicity in extinction'.

108 MacIntyre, 'Periodicity of carbonatite emplacement'.

109 Seyfert and Sirkin, *Earth History and Plate Tectonics*.

110 That a 15-million-year geomagnetic cycle exists was proposed by Mazaud et al. '15-myr periodicity'. A demonstration of a strong 30-million-year cycle is given by Pal and Creer, 'Geomagnetic reversal spurts'. The data of Pal and Creer reveal, however, that there is also a weaker interpulse in between the main spurts, yielding an overall 15-million-year periodicity. A strong-weak-strong-weak cycle with a 15-million-year periodicity is understandable in terms of the galactic theory, but not of course the Berkeley proposals of stray impacts or Nemesis. See also Negi and Tiwari, 'Matching long-term periodicities'.

111 Hoyle and Wickramasinghe, *Evolution from Space* and references therein. Hoyle and Wickramasinghe's panspermia and related claims have been subjected to severe criticism. See, for example, Davies, 'A cosmic rod in pickle' and references therein. However to date no crucial objection to the overall hypothesis has been found and their hypothesis, although circumstantial, appears to unite a remarkable number of diverse phenomena both in the sky and on the ground.

112 Alvarez et al., 'Extraterrestrial cause for the Cretaceous-Tertiary extinction'.

113 Zoller et al., 'Iridium enrichment in airborne particles'.

114 Leahy et al., 'Linking impacts and plant extinctions'; also Collinson, 'Catastrophic vegetation changes' and references therein.

115 Wolbach et al., 'Cretaceous exinctions'.

116 Bohor et al., 'Mineralogic evidence for an impact event'.

117 Hallam, 'Quaternary sea-level changes'.

118 The stray impact hypothesis was proposed for example by Gallant, *Bombarded Earth* and Urey, 'Cometary collisions with geological periods'. Gallant's pioneering work in this area has been generally neglected, perhaps because of his amateur status. However, lacking 'hard evidence' in the sky or ground, these and other early suggestions were inevitably speculative.

   Two quite distinct and parallel approaches to the catastrophism issue have been followed since the late 1970s, one geological, one astronomical. The revival of the proposal from a geological perspective was made in 1980 by Alvarez et al., 'Extraterrestrial cause', and based on the discovery of enhanced iridium at the level; and by Smit and Hertogen, 'Terrestrial catastrophe'. As described in the text, these studies have spawned many geological investigations without, however, the injection of much astronomical realism into the interpretations. Meanwhile, the authors' more general proposal based on new astronomical considerations, which included the Cretaceous-Tertiary extinctions as a particular case, had been presented in 1979: Napier and Clube, 'A theory of terrestrial catastrophism'.

119 See Clube and Napier, 'The role of episodic bombardment in geophysics', 'The microstructure of terrestrial catastrophism' and 'Giant comets and the Galaxy' and the references therein. For a convenient up-to-date summary, see Clube, *Catastrophes and Evolution*. A discussion of the effects of superwaves from ocean impacts is given in Hugget, *Cataclysms and Earth History*.

120 Corliss et al., 'The Eocene/Oligocene boundary event in the deep sea', and the references given by Van Valen, 'Catastrophes, expectations and the evidence'.

121 Jablonski, 'Background and mass extinctions'.

122 Reported in *Science News*, *129*, 356 (1986).

123 See the authors' 'Giant comets and the Galaxy'. Ideas relating to dust injection into the stratosphere have an interesting history. The astronomers Hoyle and Wickramasinghe proposed that the dinosaur extinctions may have been caused by the passage of the Earth into the dense dust tail of a comet, the blockage of sunlight cutting out photosynthesis and thus cutting the food chain at its base. In the following year Clube and Napier pointed out that large impacts were common on geological timescales and that the dust injected by the impact of a 10-km bolide would have the same effect. They therefore proposed this as one of the multifarious effects (along with blast and ozone depletion) of a bombardment episode which might lead to mass extinction (see Napier and Clube, 'A theory of terrestrial catastrophism', Clube and Napier, 'The role of episodic bombardment in geophysics', 'The microstructure of terrestrial catastrophism' and 'Giant comets and the Galaxy' and the references therein). Following this paper, the 10-km bolide impact/dust idea was taken up by Alvarez et al., 'Extraterrestrial cause', and has since been the mainstay of the 'sudden death' school of geochemists.

That a glaciation might be induced by dust injected from a meteorite or comet impact was suggested independently by the authors (see the above references; also Clube and Napier, *The Cosmic Serpent*) and by Hoyle and his colleagues (e.g. Hoyle, *Ice*). These ideas have evolved and been extended over the past few years both by Hoyle and colleagues and by the authors, and it is interesting that the role of giant comets and a fluctuating zodiacal cloud in producing glaciations has now been recognized by both groups of astronomers independently. Indeed it is possible now that the more or less continuous arrival of giant comet debris into the upper atmosphere, where it is rapidly reduced to submicron dust, plays a significant if not dominant role in the determination of climate and weather. See for example Clube and Asher, 'The Evolution of Proto-Encke'. At the time of writing these concepts have still to percolate through to the geoscience community.

That excess poisons introduced via cosmic dust may play a significant role in the extinction process has been widely recognized, arsenic (Hsu, 'Terrestrial catastrophe') and nickel (Davenport et al., 'Nickel as a toxic agent'), being just early and very recent examples of elements whose effects have been reviewed.

The role of impacts in reversing the magnetic field of the Earth, through the mechanisms discussed here, was suggested by the authors in a number of papers from 1982 onwards (e.g. Clube and Napier, 'The role of episodic bombardment in geophysics', 'The microstructure of terrestrial catastrophism' and 'Giant comets and the Galaxy'), and has since been taken up by, for example, Muller and Morris, 'Geomagnetic reversals' and Burek and Wanke, 'Impacts and glacio-eustasy'. These latter authors, however, failed to take account of the more prolonged stratospheric dusting which may take place when a very large comet enters a short-period, Earth-crossing orbit.

We emphasize that the disintegration of giant comets in the inner-planetary regions, leading to the creation of temporary zodiacal clouds and the intermittent dusting of the Earth's stratosphere, are *unavoidable* consequences of the observed astronomical environment, and that their neglect either in papers such as the above, or more generally in the scientific literature on, say, mass extinctions, can only lead to a partial or false picture of the basic mechanisms at work. Much of the current American literature dealing with catastrophism, both popular and technical, is largely erroneous because of this environmental misconception leading, as we have discussed, to a quite false dichotomy between a 'sudden death' school which sees only rare giant impacts, and a 'gradualist' school which sees no role in evolution for anything higher than the rooftops.

124 Whitmire and Jackson, 'Are periodic mass extinctions driven by a distant solar companion?'; Davis, Hut and Muller, 'Extinction of species'. Amongst the critical responses to the proposal, see Clube and Napier, 'The microstructure of terrestrial catastrophism', also 'Terrestrial catastrophism: Nemesis or Galaxy?'

125 Abt, 'The ages and dimensions of trapezium systems'.

126 Raup, 'Biological extinction in Earth history'; and Hallam, 'Asteroids and extinction'.

127 Shoemaker, 'Asteroid and comet bombardment of the Earth'.

128 Frei and Christian, *Risks of Unintentional Nuclear War*. It should be noted that in earlier authoritative discussions of the effects of nuclear weapons, major phenomena were overlooked; firstly, the catastrophic consequences of ozone depletion, and later, those of nuclear winter. Given this record, it would not be at all surprising if there is another major omission in this important area, namely the prospect of misreading one or more cosmic

impacts. To this we may add the hazard that now comes with the global distribution of nuclear power stations, any one of which is vulnerable to cosmic attack.

129 Pringle and Arkin, *SIOP*.

130 Steinbruner, 'Launch under attack'; also quoted in *Science*, 230, 156 (1986).

131 The discussion hereinafter is based on data taken from numerous nuclear war/winter studies. See for example: *The Effects on the Atmosphere of a Major Nuclear Exchange* (Anon., 1985), *Long-Term Worldwide Effects of Multiple Nuclear-Weapons Detonations* (Anon., 1975), and Dolan and Glasstone, *The Effects of Nuclear Weapons*.

132 Dansgaard et al., 'Climatic record', 'The abrupt termination'.

133 Woillard, 'Grand Pile peat bog'.

134 Clube and Napier, 'The microstructure of terrestrial catastrophism' gives references for this section.

135 The likely effects of a cosmic winter go far beyond those we have mentioned, and extensive discussion could fill another book. A popular but authoritative account of the likely effects of the analogous nuclear winter may be found in Dotto, *Planet Earth in Jeopardy*. It is pointed out therein that without modern agriculture the human population of the Earth would collapse to 1–10 per cent of its present value.

136 *Chronicon Estense*, see Ziegler, *The Black Death* for a review with references to contemporary sources.

References

Abt, H. A., 'The ages and dimensions of Trapezium systems', *Astrophysical Journal*, *304*, 688 (1986).

Alcock, L., *Arthur's Britain* (Penguin, London, 1971).

Allen, R. H., *Star Names, their Lore and Meaning* (Dover Publications Inc., New York, 1963).

Alvarez, L. W., Alvarez, W., Asaro, F. and Michel, H. V., 'Extraterrestrial cause for the Cretaceous-Tertiary extinction', *Science*, *208*, 1095 (1980).

Alvarez, W. and Muller, R. A., 'Evidence from crater ages for periodic impacts on the Earth', *Nature*, *308*, 718 (1984).

Apollodorus, tr. J. G. Frazer (Harvard University Press and Heinemann, 1979), I, 81.

Aristotle, *Meteorologica*, tr. H. D. P. Lee (Heinemann, London, 1952).

Astapovic, I. S. and Terenteva, A. K., 'Fireball radiants of the 1st–15th centuries', in *Physics and Dynamics of Meteors*, eds L. Kresak and P. Millman (Reidel, Dordrecht, 1968), 308.

Augustine, *Concerning the City of God against the Pagans*, tr. H. Bettenson (Penguin, London, 1972).

Aveni, A. S. (ed.) *World Archaeoastronomy* (Cambridge University Press, Cambridge, 1989).

Baigent, M., Leigh, R. and Lincoln, H., *The Holy Blood and the Holy Grail* (Jonathan Cape, London, 1982).

Bailey, M. E., Clube, S. V. M. and Napier, W. M., *The Origin of Comets* (Pergamon, Oxford, 1989).

Baillie, M. G. L. and Munro, M. A. R., 'Irish tree rings, Santorini and volcanic dust veils'. *Nature*, *332*, 334 (1988).

Baity, E. C., 'Archaeoastronomy and ethnoastronomy so far', *Current Anthropology*, *14* 389 (1973).

Baldwin, R. B., 'Relative and absolute ages of individual craters and the rate of infalls on the Moon in the post-Imbrium period', *Icarus*, *61*, 63 (1985).

Bellamy, H. S., *Moons, Myths and Man* (Faber & Faber, London 1936).

Bernal, J. D., *Science in History* (Watts & Co, London, 1954).

Biot, E., 'Catalogue générale des étoiles filantes et des autres météores observés en Chine', *Mémoires de l'Académie des Sciences de l'Institut National de France*, *10*, 129–352 (1807).

Blacker, C. and Loewe, M., *Ancient Cosmologies* (George Allen & Unwin, London, 1975).

Boardman, J., Griffin, J. and Murray, O., *The Oxford History of the Classical World* (Oxford University Press, Oxford, 1986).

Bohor, B. F., Foord, E. E., Modreski, P. J. and Triplehorn, D. M., 'Mineralogic evidence for an impact event at the Cretaceous-Tertiary boundary', *Science*, *224*, 867 (1984).

Bradley, R., *The Social Foundations of Prehistoric Britain* (Longman, London/New York, 1984).

Bronsted, J., *The Vikings* (Penguin, London, 1960).

Burek, P. J. and Wanke, H., 'Impacts and glacio-eustasy, plate tectonic episodes, geomagnetic reversals: a concept to facilitate detection of impact events', *Physics of Earth and Planetary Interiors*, *50*, 183 (1988).

Butterfield, H., *The Origins of History* (Methuen, London, 1981).

Callame, O. and Mulholland, J. D., 'Lunar crater Giordano Bruno: AD 1178 impact observations consistent with laser ranging results', *Science*, *199*, 875 (1978).

Clube, S. V. M. (editor), *Catastrophes and Evolution: Astronomical Foundations* (Cambridge University Press, Cambridge, 1990).

Clube, S. V. M. and Asher, D. J. 'The Evolution of Proto-Encke: Dust Bands, Close Encounters and Climatic Modulations', in *Asteroids, Comets, Meteors III* (University of Uppsala Press, 1990).

Clube, S. V. M. and Napier, W. M., *The Cosmic Serpent* (Faber & Faber, London, 1982).

Clube, S. V. M. and Napier, W. M., 'The role of episodic bombardment in geophysics', *Earth and Planetary Science Letters*, *57*, 151 (1982).

Clube, S. V. M. and Napier, W. M., 'The microstructure of terrestrial catastrophism', *Monthly Notices of the Royal Astronomical Society*, *211*, 953 (1984).

Clube, S. V. M. and Napier, W. M., 'Terrestrial catastrophism: Nemesis or Galaxy?', *Nature*, *311*, 635 and *313*, 503 (1985).

Clube, S. V. M. and Napier, W. M., 'Mankind's future, an astronomical view: ice ages, comets and catastrophes', *Interdisciplinary Science Reviews*, *11*, 236 (1986).

Clube, S. V. M. and Napier, W. M., 'Giant comets and the Galaxy: implications of the terrestrial record', in *The Galaxy and the Solar System*, eds R. Smoluchowski, J. N. Bahcall and M. S. Matthews (University of Arizona Press, Tucson, 1987).

Cohn, N., *The Pursuit of the Millennium* (Secker & Warburg, London, 1957).

Collinson, M. E., 'Catastrophic vegetation changes', *Nature*, *324*, 112 (1986).

Corliss, B. H., Aubrey, M. P., Berggren, W. A., Fenner, J. M., Keigwin, L. D. and Keller, G., 'The Eocene/Oligocene boundary event in the deep sea', *Science*, *226*, 806 (1984).

Cornford, F. M., *Principium Sapientiae* (Peter Smith, Gloucester, Mass., 1971).

Creer, K. M. and Pal, P. C., 'On the frequency of reversals of the geomagnetic dipole', in *Catastrophes and Evolution: Astronomical Foundations*, ed. Clube, S. V. M. (Cambridge University Press, Cambridge, 1989), 113.

Cronin, V., *The View from Planet Earth* (Collins, London, 1981).

Dall'Olmo, U. 'Latin terminology relating to aurorae, comets, meteors and novae', *Journal of the History of Astronomy*, *11*, 10 (1980).

Dansgaard, W., Johnsen, S. J., Clausen, H. B. and Langway, C. C., 'Climatic record revealed by the Camp Century ice core', in *The Late Cenzoic Glacial Ages*, ed. K. K. Turekian (Yale University Press, Newhaven, Conn., 1971), 35.

Dansgaard, W., White, J. W. C. and Johnsen, S. J., 'The abrupt termination of the Younger Dryas climate event', *Nature*, *339*, 532 (1989).

Darlington, C. D., *The Evolution of Man and Society* (George Allen & Unwin, London, 1969).

Davenport, S., Wdowiak, T. J., Jones, W. and Wdowiak, P., 'Nickel as a toxic agent in bolide impacts', Geological Society of America, special paper. In press (1989).

David, R., *Cult of the Sun* (Dent, London, 1980).

Davidson, M., *The Stars and the Mind* (Scientific Book Club, London, 1947).

Davies, R. E., 'A cosmic rod in pickle', *Nature*, *324*, 10 (1986).

Davis, M., Hut, P. and Muller, R. A., 'Extinction of species by periodic comet showers', *Nature*, *308*, 715 (1984).

Deacon, E. L., 'The 1908 Tunguska explosion', *Weather*, *37*, 6 (1982).

de Laubenfels, M. W., 'Dinosaur extinction: one more hypothesis', *Journal of Paleontology* 30, 207 (1956).

Dicks, D. R., *Early Greek Astronomy to Aristotle* (Thames & Hudson, London, 1970).

Dolan, S. and Glasstone, P. J., *The Effects of Nuclear Weapons*, 3rd edn (US Government Printing Office, Washington DC, 1977).

Dorman, F. H., 'Some Australian oxygen isotope temperatures and a theory for a 30-million-year world-temperature cycle', *Journal of Geology*, *76*, 297 (1968).

Dotto, L., *Planet Earth in Jeopardy* (Wiley, Chichester, 1986).

Farrington, B., *Greek Science* (Penguin, London, 1944).

Fischer, A. G. and Arthur, M. A., 'Secular variations in the pelagic realm', *Soc. Econ. Paleont. Mineral*, special publication, *25*, 19 (1977).

Fisher, D. J. V., *The Anglo-Saxon Age* (Longman, London, 1973).

Flammarion, G. C. and Danjon, A., eds, *The Flammarion Book of Astronomy* (George Allen & Unwin, London, 1968).

Folliot, K. A., *Atlantis Revisited* (Professional Books, Richmond, 1984).

Forsyth, N., *The Old Enemy* (Princeton University Press, Princeton, NJ, 1987).

Frankfort, H., *Kingship and the Gods* (University of Chicago Press, Chicago, 1948).

Frankfort, H. and Frankfort, H. A., 'Myth and reality', in *The Intellectual Adventure of Ancient Man*, eds H. and H. A. Frankfort (University of Chicago Press, Chicago, 1946).

Frazer, J. G., *The Golden Bough* (Macmillan Press, London, 1978).

Frei, D. and Christian, C., *Risks of Unintentional Nuclear War* (Croom Helm, Beckenham, Kent, 1983).

Fuhr, I., *Ein Altorientalisches Symbol* (Harrassowitz, Wiesbaden, 1967), in German.

Gallant, R. L., *Bombarded Earth* (John Baker, London, 1964).

Gardiner, Sir A., *Egypt of the Pharaohs* (Oxford University Press, Oxford, 1961).

Gibbon, E., *The Decline and Fall of the Roman Empire*, abridged by D. M. Low (Guild Publishing, London, 1987).

Goudie, A., *Environmental Change* (Clarendon Press, Oxford, 1983).

Graves, R., *Greek Myths I, II* (Penguin, London, 1955).

Grove, J., *The Little Ice Age* (Methuen, London, 1987).

Guillemin, A., *Les Etoiles Filantes* (Libraire Hachette, Paris, 1889).

Guthrie, W. K. C. *Orpheus and Greek Religion* (Methuen, London, 1935).

Hadingham, E., *Circles and Standing Stones* (Anchor Press/Doubleday, New York, 1976).

Hallam, A., 'Asteroids and extinction: no cause for concern', *New Scientist* (8 November 1984), 30.

Hallam, A., 'Quaternary sea-level changes', *Annual Review of the Earth and Planetary Sciences*, *12*, 205 (1984).

Hart, M. H., 'The evolution of the atmosphere of the Earth', *Icarus*, *33*, 23 (1978).

Hartung, J. B., 'Was the formation of a 20-km-diameter impact crater on the Moon observed on June 18, 1178?', *Meteoritics*, *11*, 187 (1976).

Heath, T. L., *Aristarchus of Samos* (Clarendon Press, Oxford, 1913).

Heggie, D. C., *Megalithic Science* (Thames & Hudson, London, 1981).

Hellman, C. D., *The Comet of 1577: its Place in the History of Astronomy* (Columbia University Press, New York, 1944).

Herodotus, *The Histories*, tr. A. de Selincourt (Penguin, London, 1954).

Hetherington, N. S. and Ronan, C. A., 'Ptolemy's Almagest: fourteen centuries of neglect', *Journal of the British Astronomical Association*, *94*, 256 (1984).

Ho, Ping-Yü, 'Ancient and medieval observations of comets and novae in Chinese sources', *Vistas in Astronomy*, *5*, 127 (1962).

Holmes, A., *The Age of the Earth* (Benn, London, 1927).

Homer, *The Iliad*, tr. E. V. Rieu (Penguin, London, 1950).

Hoyle, F., *Ice* (Dent, London, 1982).

Hoyle, F. and Wickramasinghe, N. C., *Evolution from Space* (Dent, London, 1981).

Hsu, K. J., 'Terrestrial catastrophe caused by cometary impact at the end of the Cretaceous', *Nature*, *285*, 201 (1980).

Huggett, R. J., *Cataclysms and Earth History: the Development of Diluvialism* (Clarendon Press, Oxford, 1990).

Huxley, F., *The Way of the Sacred* (W. H. Allen, London, 1980).

Jablonski, D., 'Background and mass extinctions: the alternation of macro-evolutionary regimes', *Science*, *231*, 129 (1986).

Jacobsen, T., 'Mesopotamia', in *The Intellectual Adventure of Ancient Man*, eds H. and H. A. Frankfort (University of Chicago Press, Chicago, 1946).

Jacobsen, T., *The Treasures of Darkness* (Yale University Press, Newhaven, Conn., 1978).

Jaki, S. L., *The Milky Way* (David & Charles, Newton Abbot, 1973).

James, E. O., *The Ancient Gods* (Weidenfeld & Nicolson, London, 1962).

James, E. O., *The Worship of the Sky God* (Athlone Press, London, 1963).

Keynes, G., *Newton, the Man*, Royal Society Newton Tercentenary Proceedings, 15–19 July 1946 (Cambridge University Press, Cambridge, 1947), 27–34.

King, H. C., *The Background of Astronomy* (Watts, London, 1957).

Kitto, H. D. F., *The Greeks* (Penguin, London, 1951).

Klinkerfues, W., 'Gottinger Nachtrichten' (1873) 275; tr. W. J. Fisher in *Popular Astronomy*, *39*, 573 (1931).

Kresak, L., 'The Tunguska object: a fragment of Comet Encke?', *Bulletin of the Astronomical Institute of Czechoslovakia*, *29*, 129 (1978).

Kresak, L., 'Sources of interplanetary dust', in *Solid Particles in the Solar System*, eds. I. Halliday and B. A. McIntosh (Reidel, Dordrecht, 1980), 211.

Krinov, E. L., *Principles of Meteoritics*, tr. J. S. Romankiewicz (Pergamon, Oxford, 1960).

Kronk, G. W., *Meteor Showers* (Enslow, Aldershot, 1988).

Krupp, E. C., ed., *In Search of Ancient Astronomies* (Chatto & Windus, London, 1979).

Krupp, E. C., 'The "binding of the years", the Pleiades, and the nadir sun', *Archaeoastronomy*, *5*, 10 (1982).

Krupp, E. C., *Echoes of the Ancient Skies* (Harper & Row, New York, 1983).

Krupp, E. C., ed., 'Archaeoastronomy and the roots of science', AAAS Selected Symposium 71 (Westview Press, Boulder, 1984).

Kuhn, T., *The Structure of Scientific Revolutions*, 2nd edn (University of Chicago Press, Chicago, 1962).

Lamb, H. H., *Climate: Present, Past and Future* (Methuen, London, 1977).

Leahy, G. D., Spoon, M. D. and Retallack, G. J., 'Linking impacts and plant extinctions', *Nature*, *318* (1985).

Leggett, J. K., McKerrow, W. S., Cocks, L. R. M. and Rickard, R. B., 'Periodicity in the Lower Paleozoic marine realm', *Journal of the Geological Society*, *138*, 167 (1981).

Levin, B. Y. and Bronshten, V. A., 'The Tunguska event and the meteors with terminal flares', *Meteoritics*, *21*, 199 (1986).

Lucretius, *De Rerum Natura*, tr. R. E. Latham (Penguin, London, 1951).

McCrea, W. H., 'Long time-scale fluctuations in the evolution of the Earth', *Proceedings of the Royal Society*, A *375*, 1 (1981).

MacIntyre, R. M., 'Periodicity of carbonatite emplacement', *Nature*, *230*, 79 (1971).

Mackenzie, D. A., *Indian Myth and Legend* (Gresham Publishing Co., London, 1913).

McKie, E., *The Megalith Builders* (Phaidon, London, 1977).

Mazaud, A., Laj, C., de Sèze, L. and Verosub, K. B., '15-Myr periodicity in the frequency of geomagnetic reversals since 100 Myr', *Nature*, *304*, 328 (1983).

Mellart, J., 'Egyptian and Near Eastern Chronology', *Antiquity*, *53*, 6, 1979.

Milbrath, S., 'Star gods and astronomy of the Aztecs', in *La Antropologia Americanista en la Actualidad*, eds Mexicanos Unidos, *1*, 289 (1980).

Monmouth, Geoffrey of, *History of the Kings of Britain*, tr. L. Thorpe (Penguin, London, 1966).

Morris, C., *Historical Tales: the Romance of Reality* (Lippincott, 1893).

Morris, J., *The Age of Arthur* (Phillimore, Chichester, 1977).

Morris, J., *Nennius* (Phillimore, Chichester, 1980).

Muller, R. A. and Morris, D. E., 'Geomagnetic reversals from impacts on the Earth', *Geophysical Research Letters*, *13*, 1177 (1986).

Myres, J. N. L., *The English Settlements* (Clarendon Press, Oxford, 1986).

Napier, W. M. and Clube, S. V. M., 'A theory of terrestrial catastrophism', *Nature*, *282*, 455 (1979).

Needham, J., *The Grand Titration* (George Allen & Unwin, London, 1979).

Negi, J. G. and Tiwari, R. K., 'Matching long-term periodicities of geo-magnetic reversals and galactic motions of the Solar System', *Geophysical Research Letters*, *10*, 713 (1983).

Neugebauer, O., 'The history of ancient astronomy: problems and methods', *Journal of Near Eastern Studies*, *4*, 1 (1945).

New Larousse Encyclopedia of Mythology (Hamlyn, London, 1959).

Newton, R. R., *The Crime of Claudius Ptolemy* (Johns Hopkins University Press, Baltimore, 1977).

Nonnos, *Dionisiaca*, tr. W. H. D. Rouse (Heinemann, London, 1940).

Oates, J., *Babylon*, revised edn (Thames & Hudson, London, 1986).

Olsson-Steel, D., 'Asteroid 5025 P-L, Comet 1967 II Rudnicki and the Taurid Meteoroid Complex', *Observatory*, *107*, 157 (1987).

Opik, E. J., 'On the catastrophic effects of collisions with celestial bodies', *Irish Astronomy Journal*, *5*, 34 (1958).

Ovid, *Metamorphoses*, tr. and ed. F. J. Miller, 2nd edn (Heinemann, London, 1984).

Pal, P. C. and Creer, K. M., 'Geomagnetic reversal spurts and episodes of extraterrestrial catastrophism', *Nature*, *320*, 148 (1986).

Plato, *Timaeus and Critias*, tr. H. D. P. Lee (Penguin, London, 1965).

Pliny, *Natural History II*, tr. H. Rackham (Heinemann, London, 1968).

Plutarch, *Concerning Isis and Osiris*, tr. B. Perrin (Heinemann, London, 1982).

Pringle, P. and Arkin, W., SIOP (Sphere, London, 1983).

Pritchard, J. B., *The Ancient Near East. An Anthology of Texts and Pictures, I, II* (Princeton University Press, Princeton, 1958).

Ptolemy, *The Almagest*, tr. G. J. Toomer (Duckworth, London, 1984).

Ptolemy, *The Tetrabiblos*, tr. F. E. Robbins (Heinemann, London, 1980).

Rampino, M. R. and Stothers, R. B., 'Geological rhythms and cometary impacts', *Science*, *226*, 1427 (1984).

Raup, D. M., 'Biological extinction in Earth history', *Science*, *231*, 1528 (1986).

Raup, D. M. and Sepkoski, J. J., 'Periodicity of extinctions in the geologic past', *Proceedings of the National Academy of Sciences of the USA*, *81*, 801 (1984).

Reiche, H. A. T., 'The language of archaic astronomy: a clue to the Atlantis myth?', *Technology Review* (December 1977), 85.

Renfrew, C., *Before Civilization* (Penguin, London, 1970).

Rhys, J., *Celtic Heathendom* (London, 1888).

Riley-Smith, J., *The First Crusade and the Idea of Crusading* (Athlone Press, London, 1986).

Rudaux, L. and de Vaucouleurs, G., *Larousse Encyclopedia of Astronomy* (Batchworth, London, 1959).

Rundle-Clark, R. T., *Myth and Symbol in Ancient Egypt* (Thames & Hudson, London, 1959).

Russell, C. T., Aroian, R., Arghavani, M. and Nock, K., 'Interplanetary magnetic field enhancements and their association with the asteroid 2201 Oljato', *Science*, *226*, 43 (1984).

Salway, P., *Roman Britain* (Clarendon Press, Oxford 1981).

Sandars, N. K., *The Sea Peoples* (Thames & Hudson, London, 1985).

Santillana, G. de and Dechend, H. von, *Hamlet's Mill* (Gambit Press, Boston 1969).

Sanderson, R., 'The night it rained fire', *Griffith Observer* (November 1984), 2.

Savage, A., *The Anglo-Saxon Chronicles* (Macmillan, London, 1982).

Schafer, E. H., *Pacing the Void* (University of California Press, California, 1977).

Sekanina, Z., 'The Tunguska event: no cometary signature in evidence', *Astronomical Journal*, *88*, 1382 (1983).

Seneca, *Naturales Quaestiones*, tr. E. H. Corcoran (Heinemann, London, 1972).

Sepkoski, J. J., 'Periodicity in extinction and the problem of catastrophism in the history of life', *Journal of the Geological Society*, *146*, 7 (1989).

Seyfert, C. K. and Sirkin, L. A., *Earth History and Plate Tectonics* (Harper & Row, New York, 1979).

Shackleton, N. J., 'Climate and extinctions in the deep-sea sediment record', in *Evolution and Extinction*, Discussion Meeting of the Royal Society, 9–10 November 1988. In press.

Shoemaker, E. M., 'Asteroid and comet bombardment of the Earth', *Annual Review of the Earth and Planetary Sciences*, *11*, 461 (1983).

Simmons, I. G. and Tooley, M. J., *The Environment of British Prehistory* (Duckworth, London, 1981).

Smart, N., *The Religious Experience of Mankind* (Fontana, London, 1971).

Smit, J. and Hertogen, J., 'Terrestrial catastrophe caused by cometary impact at the end of Cretaceous', *Nature*, *285*, 201 (1980).

Snyder, G. S., *Maps of the Heavens* (Andre Deutsch, London, 1984).

Southern, R. W., *Western Society and the Church in the Middle Ages* (Penguin, London, 1970).

Stecchini, L. C., 'The Newton affair', *Kronos*, *9*, 34; *9*, 52; *10*, 62 (1984).

Steinbruner, J., 'Launch under attack', *Scientific American*, *250*, 23 (1984).

Steiner, J. and Grillmair, E., 'Possible galactic causes for periodic and episodic glaciations', *Geological Society of America Bulletin*, *84*, 1003 (1973).

Stohl, J., 'On the distribution of sporadic meteor orbits', in *Asteroids, Comets, Meteors, I*, eds C. I. Lagerkvist and H. Rickman (Uppsala University, Sweden, 1983) 565.

Sybilline Oracles, in the *Encyclopedia of Religion and Ethics II*, eds M. Hastings and E. Selbie (T & T Clark, Edinburgh, 1908–26).

Thomas, K., *Religion and the Decline of Magic* (Weidenfeld & Nicolson, London, 1971).

Thompson, L. G. and Mosley-Thompson, E., 'Microparticle concentration variations linked with climatic change: evidence from polar ice cores', *Science*, *212*, 812 (1981).

Tian-shan, Z., 'Ancient Chinese observations of meteor showers', *Chinese Astronomy I* (1977), 197, entry no. 39.

Tolstoy, N., *The Quest for Merlin* (Hamish Hamilton, London, 1985).

Urey, H. C., 'Cometary collisions with geological periods', *Nature*, *242*, 32 (1973).

Van der Waerden, B. L., *Science Awakening II: the Birth of Astronomy* (Noordhof International Publishers, Leiden, 1974).

Van Valen, L. M., 'Catastrophes, expectations and the evidence', *Palaeontology*, *10*, 121 (1984).

Velikovsky, I., *Worlds in Collision* (Victor Gollancz, London 1950).

Velikovsky, I., *Mankind in Amnesia* (Sidgwick & Jackson, London, 1982).

Virgil, *The Aeneid*, tr. H. R. Fairclough (Heinemann, London, 1978).

Waddell, W. G., *Manetho* (Heinemann, London, 1978).

Walcot, P., *Hesiod and the Near East* (University of Wales Press, Cardiff, 1966).

West, M. L., *Early Greek Philosophy and the Orient* (Oxford University Press, Oxford, 1971).

Whipple, F. L., *The Mystery of the Comets* (Cambridge University Press, Cambridge, 1985).

Whipple, F. L. and Hamid, S. E., 'On the origin of the Taurid meteor streams', *Helwan Observatory Bulletin*, *41*, 224 (1952).

Whiteside, D. T., 'Before the Principia: the maturing of Newton's thoughts on dynamical astronomy, 1664–1684', *Journal for the History of Astronomy*, *1*, 5 (1970).

Whitlock, R., *In Search of Lost Gods* (Phaidon, London, 1979).

Whitmire, D. P. and Jackson, A. A., 'Are periodic mass extinctions driven by a distant solar companion?', *Nature*, *308*, 713 (1984).

Wilson, J. A., 'Egypt', in *The Intellectual Adventure of Ancient Man*, eds H. and H. A. Frankfort (University of Chicago Press, Chicago, 1946).

Winterbottom, M., *Gildas* (Phillimore, Chichester, 1978).

Woillard, G., 'Grande Pile peat bog: a continuous pollen record for the last 140,000 years', *Quaternary Research*, *9*, 1 (1978).

Wolbach, W. S., Lewis, R. S. and Anders, E., 'Cretaceous extinctions: evidence for wildfires and search for meteoritic material', *Science*, *230*, 167 (1985); *234*, 261 (1985).

Wood, M., *In Search of the Trojan War* (BBC Publications, London, 1985).

Yeomans, D. K. and Kiang, T., 'The long-term motion of Halley's Comet', *Monthly Notices of the Royal Astronomical Society*, *197*, 633 (1981).

Ziegler, P., *The Black Death* (Penguin, London, 1982).

Zoller, W. H., Parrington, J. R. and Phelan Kotra, J. M., 'Iridium enrichment in airborne particles from Kilauea volcano: January 1983', *Science*, *222*, 1118 (1983).

Index